Asymmetry Indexes, Behavioural Instability and the Characterization of Behavioural Patterns

Asymmetry Indexes, Behavioural Instability and the Characterization of Behavioural Patterns

Special Issue Editor

Cino Pertoldi

MDPI • Basel • Beijing • Wuhan • Barcelona • Belgrade • Manchester • Tokyo • Cluj • Tianjin

Special Issue Editor
Cino Pertoldi
Department of Chemistry and
Bioscience, Aalborg University
Denmark

Editorial Office
MDPI
St. Alban-Anlage 66
4052 Basel, Switzerland

This is a reprint of articles from the Special Issue published online in the open access journal *Symmetry* (ISSN 2073-8994) (available at: https://www.mdpi.com/journal/symmetry/special_issues/Asymmetry_Indexes_Behavioural_Instability_Characterization_Behavioural_Patterns).

For citation purposes, cite each article independently as indicated on the article page online and as indicated below:

LastName, A.A.; LastName, B.B.; LastName, C.C. Article Title. *Journal Name* **Year**, *Article Number*, Page Range.

ISBN 978-3-03936-056-7 (Pbk)
ISBN 978-3-03936-057-4 (PDF)

Cover image courtesy of Anika Gottschalk, Henriette Lyhne and Anne Cathrine Linder.

Contents

About the Special Issue Editor

Cino Pertoldi obtained his Master degree in Natural Sciences from the University of Milan (Italy) and his PhD in Conservation Biology from the University of Aarhus (Denmark). After several stays in Spain and Poland, he worked at the Danish Ministry of Environment Research Institute and spent several years as Associate Professor at the University of Aarhus. He was EU Professor at the Mammal Research Institute in Bialowieza (Poland) and is now Full Professor at the University of Aalborg Department of Chemistry and Bioscience and Aalborg Zoo. His research focuses on the empirical conservation and evolutionary genetics of animals, but also includes conceptual and theoretical studies in the interface between genetics, ecology, and evolution. He has merged current efforts in evolutionary and ecological genetics, complementing molecular genomics and macroecology in order to understand how genetic measures can indicate causal processes. His interdisciplinary approach includes experimental, theoretical/computational, and empirical approaches which allow a holistic vision of the dynamics of natural processes.

Editorial

EDITORIAL: Asymmetry Indexes, Behavioral Instability and the Characterization of Behavioral Patterns

Cino Pertoldi [1,2,*], Sussie Pagh [1] and Lars Arve Bach [1]

[1] Department of Chemistry and Bioscience, Aalborg University, Fredrik Bajers Vej 7H, DK-9000 Aalborg, Denmark
[2] Department of Zoology, Aalborg Zoo, Mølleparkvej 63, DK-9000 Aalborg, Denmark
* Correspondence: cp@bio.aau.dk

Received: 20 April 2020; Accepted: 21 April 2020; Published: 24 April 2020

A change in a behavior is often the first and fast reaction to an environmental (external) or physiological (internal) *stimulus* that animals (and plants) are exposed to. Behavioral responses are thus important for the ability of organisms to survive and reproduce in constantly changing environments. Differences between individuals in behavior are due to differences in the environmental *stimuli* that they are experiencing or have experienced and their interactions with the genetic profile of the individuals, which in turn can affect the way in which individuals experience the environment.

Several behavioral studies have been conducted disregarding the individuality of the behavioral phenotypes (sometimes referred to as personalities) of animals. This can reduce the reproducibility of the results and lead to the incorrect conclusion that behavioral changes are random processes. An insufficient understanding of behavioral changes is also caused by the unpredictability of animal behavior, i.e., behavioral instability. Behavioral instability is subsequently a behavioral component that should be incorporated into behavioral studies.

Behavioral studies typically describe behavioral traits in (i) a qualitative way (presence/absence); (ii) a semiquantitative way (minor, medium, and maximum expression of a certain trait); or (iii) fully quantify the behavior by measuring, for example, the speed and the distance travelled by an individual or by estimating the frequency at which a given behavior is occurring during a specified time interval. The time interval is sometimes randomly chosen, and in this way the interdependence of the behaviors is neglected.

There is an urgent need for more quantitative studies covering large periods of observations, and there is also a need for a standardized statistical pipeline and ways of presenting behavioral patterns, in order to make different studies comparable. Statistical procedures should include methods that avoid the reduction of the variation of a given parameter. This can be achieved by using suitable statistical transformations, in order to make the distributions of the data normally distributed and to homogenize the variances as much as possible (assumptions required by several parametric tests). Such quantitative studies will generate important information about the variability (due to different personality of the individuals) and/or the predictability of behavioral traits that are currently often ascribed to inconvenient noise, i.e., variation of behavioral traits is often considered random noise and the extreme values are typically considered outliers and therefore removed.

For the above-mentioned reasons, we propose a change of research direction that will complement the classical approaches so far used in behavioral studies. We propose a holistic concept of behavioral instability that comprehends a series of parameters that describe the variation of a distribution using modified indices that are traditionally used to investigate developmental instability (like fluctuating asymmetry, asymmetry index, and directional asymmetry).

Behavioral instability can be utilised for described behavioral traits like, for example, changes in directions during a movement, or the time spent on a certain activity, but can also be applied to physiological measures and to molecular and cellular mechanisms (if they are quantifiable e.g., in terms of duration and/or intensity of a process).

Behavioral instability can be described in terms of time, i.e., the distribution of the time-intervals in which a given behavior occurs. Moreover, it can be described in terms of spatial distributions, i.e., how individuals are distributed in a population/patch, or can be described in terms of binary distribution (Bernoulli distribution) where the frequency of two binary *status* of a behavior can be quantified.

When distributions of a trait are obtained from investigations, the four moments of distributions (mean, variance, skewness, and kurtosis) and some modified measurements (depending on the kind of distribution and its characteristics) can be utilised for describing the behavioral characteristics. In particular, variance, skewness, and kurtosis can provide accurate estimates of the probability that a given behavior will occur and with which intensity. In addition, the analysis of the distribution can give indications about the heterogeneity of the individuals and can allow an estimation of the number of different personalities present in a group of individuals. This analysis, called "admixture analysis", can provide important information about the population's capacity to adapt in a plastic way through behavioral means. The presence of different personalities in a population is comparable to genetic variability in the population, and hence higher variability can be translated into higher capacity or higher resilience of the population versus sudden and unpredictable environmental changes.

The individual personality can be shown by behavioral reaction norms, and the concept of behavioral instability can be applied. This method can provide researchers with a relatively unbiased assessment of behavioral responses, thus enabling the reproducibility of results.

The fact that behavioral instability does not need bilateral traits to estimate instability is clearly expanding its scope for different applications.

Several previous behavioral studies have been conducted disregarding the personalities of the animals. This approach has considerably reduced the reproducibility of the results, thus causing the misapprehension of the conclusion that behavioral changes are random processes.

The novel concept of behavioral instability presents new perspectives in the field of quantitative genetics and in associated fields. Studying behavioral traits using the suggested approaches could have significant potential in evolutionary studies to evaluate, e.g., the plasticity and genotypic difference between individuals and in psychological human studies.

Several techniques such as proteomic tools and next-generation sequencing have been applied with the attempt to discover the molecular and cellular mechanisms of phenotypic plasticity and canalization. Similarly, several genome-wide association studies are trying to associate genetic variation with variation in behavioral traits. These studies will clearly be beneficial for future research given the potential to associate the concept of behavioral instability with genetic variation in order to estimate the heritability of the different aspects of behavioral instability.

Article

Using Behavioral Instability to Investigate Behavioral Reaction Norms in Captive Animals: Theoretical Implications and Future Perspectives

Anne Cathrine Linder [1,†], Anika Gottschalk [1,†], Henriette Lyhne [1,†], Marie Gade Langbak [1], Trine Hammer Jensen [2] and Cino Pertoldi [1,2,*]

[1] Department of Chemistry and Bioscience, Aalborg University, Fredrik Bajers Vej 7H, DK-9000 Aalborg, Denmark; c.linder04@gmail.com (A.C.L.); anika@gottschalk-buxtehude.de (A.G.); henry.lyhne@hotmail.com (H.L.); marielangbak@gmail.com (M.G.L.)
[2] Department of Zoology, Aalborg Zoo, Mølleparkvej 63, DK-9000 Aalborg, Denmark; trine@bio.aau.dk
* Correspondence: cp@bio.aau.dk
† The three authors contributed equally.

Received: 9 March 2020; Accepted: 2 April 2020; Published: 10 April 2020

Abstract: Behavioral instability is a concept used for indicating environmental stress based on behavioral traits. This study investigates the possibility of using behavioral instability as a tool for assessing behavioral reaction norms in captive animals. The understanding of personality in captive animals can be a useful tool in the development of enrichment programs in order to improve animal welfare. In this study, a case study examined how olfactory stimuli affected the behavior of two polar bears *Ursus maritimus* in captivity. Using continuous focal sampling throughout the day, it was found that for many behaviors, the individuals responded differently to stimuli, indicating that there was a difference in behavioral reaction norms. This is shown using multiple approaches. One approach used traditional methods for behavioral analyses, and the other approach used the concept of behavioral instability as a new quantitative method. This study demonstrates the utility of behavioral instability as a new quantitative method for investigating behavioral reaction norms, expanding the possibility of comparing behavioral responses between species. Moreover, it is shown that outliers—that cause asymmetric distributions—should not be removed in behavioral analysis, without careful consideration. In conclusion, the theoretical implications and future perspectives of behavioral instability are discussed.

Keywords: asymmetry; ethogram; olfactory stimuli; stereotypy; *Ursus maritimus*; enrichment; captivity; asymmetric diversity

1. Introduction

It has been shown for several species that conspecifics have different behavioral reaction norms [1–3]. These different behavioral reaction norms are expressed by consistent behavioral responses under various conditions that can vary in different ways, for example, population density, stress and enrichment [2,4,5]. A behavioral reaction norm is a set of behavioral phenotypes that a single individual produces in a specified set of environments [6]. The behavioral responses of an animal can influence its welfare, as these responses can vary between individuals; that is, an environmental condition may be well tolerated by one individual, but not by another [7]. Stereotypic behavior is described as a repetitive motion with no apparent purpose and has generally been shown to be a sign of stress, due to its correlation with increased corticoid levels, thus making stereotypy an indication of poor welfare [8,9].

Several studies have shown that enrichment and the presence of choice in activity is negatively correlated with stereotypy [10–14]. Carlstead and Seidensticker [11] concluded that an olfactory stimulus, at least during breeding season, was sufficient to distract the bear from pacing. However, other studies have shown that not all enrichments improve welfare when measured in time spent on stereotypy [15]. This could be explained by the variation in the tested individuals' behavioral responses [7]. To improve the welfare of polar bears and other large predatory animals in captivity, it would be relevant to quantify their behavior and behavioral reaction norms in order to understand how their general welfare and the welfare of each individual can be improved [9]. It is important to investigate whether different animals have different behavioral reaction norms, as they would be expected to react differently to stimuli, either increasing or decreasing their time spent on stereotypic behavior, also leading to a difference in welfare. Rose et al. [16] and Shyne [13] emphasize the need for further development of quantitative assessments of animal welfare in order to increase the reliability of non-invasive welfare indicators, such as behavioral traits.

The sampling methods used in the traditional studies of animal behavior vary between studies and have been described and compared in Altmann [17]. Bashaw et al. [18] found that there was a difference in behavior throughout the day, suggesting that the assessment should be carried out not only at a specific time of the day, but for a longer period of time, covering a larger proportion of the day. Standardizing these sampling methods would contribute to a quantitative and systematic behavior analysis.

Different suggestions have been made to improve the traditional non-standardized methods using ethograms and observations of different time intervals, by using more quantitative and systematic methods. Pertoldi et al. [19] introduced the concept of behavioral instability based on the concept of developmental instability. Behavioral instability was introduced as a method of studying the symmetry of behavior, by observing bilateral behavioral traits, for example, how many times an individual looks to the left, versus the right or up and down. Bech-Hansen et al. [20] introduced two variables to this concept, BSYM and BVAR. BSYM is the behavioral instability of symmetry, meaning the deviation from a symmetric distribution for the studied behavior; BVAR is the variance of residuals for the studied behavior, where a higher variance indicates a smaller capacity for anticipating a behavior when stressors are present [19]. The concept of behavioral instability could, as proposed by Bech-Hansen et al. [19], also be applied to measure the effect of environmental stress on behavioral data other than bilateral data, as it can be used to measure the effect of environmental stress. Therefore, whether behavioral instability can be used as a new, quantitative way of studying behavior and behavioral responses should be investigated.

Aim of the Paper

This study aims to investigate the application of the concept of behavioral instability as a tool for studying the behavioral responses of captive animals and to provide a theoretical framework and a statistical pipeline for the analysis of the data. This will be achieved through a case study that investigates the behavioral reaction norms of polar bears in captivity by comparing the effect of olfactory stimuli on two individuals at Aalborg Zoo, Denmark. It was anticipated that the stimuli would have an effect on the individuals' behavior and that there would be a difference between the two individuals' behavioral reaction norms, thus enabling the investigation of behavioral instability to quantify this difference in behavioral responses. Here it was expected that behavioral instability can be utilized as a tool for quantifying the differences in animal behavior and therefore applicable as a new method for studying animal behavioral reaction norms.

2. Methods

2.1. Animals and Setting

In the case study, the behavior of two female polar bears at Aalborg Zoo in Denmark was observed. The two individuals are siblings that were born in November 2016 at Aalborg Zoo. The sisters have been kept in a separate enclosure from their mother since spring 2019. The two enclosures were separated by a dry moat, giving the two individuals visual access to their mother. Their diet consisted of vegetables, fruit, fish, meat (primarily horse intestines), dog kibble and various treats such as dried dates, which they were fed randomly throughout the week. The area of the enclosure used for this study was 768 m^2 and consisted of a pool, land covered by gravel and concrete and a den (a map of the enclosure can be seen in Appendix A). The windows for the zoo visitors were placed opposite the den, making the inside of the den not visible to visitors. The zookeepers were able to access the polar bears when they were in the den; this is also where the zookeepers would occasionally train the polar bears and feed them treats.

2.2. Data Collection

The observations took place from the beginning of October to the beginning of November 2019 during the zoo's off-season. Nine observation sessions were spread throughout this time period. The observation sessions were conducted by filming the polar bears using four action cameras (Kitvision Escape HD 5) that were placed around the enclosure, ensuring video surveillance of the entire outdoor perimeter (camera placement can be seen in Appendix A). Each session began at sunrise, ranging from 07:29 (UTC+2) to 08:34 (UTC+1) and lasted for nine hours. Three of the observation sessions were control treatments (treatment C), which were used as a baseline measurement of the polar bears' behavior under normal conditions. During three of the other observation sessions, the bears were given stimuli in the form of two dog-scented objects (treatment D), one for each individual, which were thrown into the enclosure between 09:00 and 09:30 and left in the enclosure for the remainder of the observation session. Each dog-scented object was scented by a different dog, thus two dogs contributed with their scents for each observation session. This choice of enrichment is based on the observations of the zookeepers, as they have noticed that the two polar bears are especially reactive when dogs are among the zoo visitors. The objects were fabric boxes that were placed in the beds of different dogs for approximately a week prior to each of the three observation sessions, thus scenting the boxes with the natural odor of the dogs. For each observation session new fabric boxes were used, thus ensuring the confounding factor of the novel scent receptacle, as the scent does not accumulate [21]. In order to estimate the effect of the dog odor and not the novelty of the object itself, three observation sessions were used to observe the effect of the unscented fabric boxes. The behavioral data for the observation sessions with unscented objects were only used to confirm that the effect of the stimuli came from the dog odors and not the fabric boxes themselves; these data were only used in a preliminary analysis. The preliminary analysis of the individuals' behavior when exposed to the unscented boxes, showed a slight deviation in their behavior compared to treatment C, whereas a larger difference was found when compared to treatment D. Hence, indicating that the effect resulted primarily from the olfactory stimuli and not from the novelty of the object itself.

2.3. Analysis

Behavioral observations were based on the analysis of the filmed material by four coders, using the ethogram described in Table 1. Interaction with the object in treatment D was accounted for as part of the behaviors: 'activity on land', 'activity in water' and 'social play'. Prior to this, a concordance test (≥85%) was performed to ensure that the inspections of all four coders were in agreement. The footage was analyzed using continuous focal sampling of the nine hours that each observation session lasted [17]. Furthermore, all occurrences were treated as states as described by Altmann [17]; thus, for each observation session, all 32,400 seconds were coded for each individual. The preliminary analysis was based on all nine observation sessions, amounting to 583,200 seconds and 3322 data points. Further analyses were based on only six observation sessions, three for treatment C and three for treatment D, amounting to 388,800 seconds and 2236 data points.

Table 1. Behavioral ethogram.

Behavior	Description
Activity on land	Locomotion and interaction with objects while on land
Activity in water	Locomotion and interaction with objects while submerged in water
Social play	Individuals interacting playfully of fighting with each other, possibly while interacting with objects.
Stereotypic	Repeating a specific walking pattern or movement aimlessly
Inactive	Resting or sleeping; laying down or sitting with minimal movement
Inside	Inside the den and therefore out of sight
Other	Eating, drinking, urinating, defecating, maintenance of coat (e.g., by rolling in gravel) and out of sight due to blind camera angles

The statistical analyses were conducted in RStudio version 3.6.0 [22] and Past version 3.26b [23]. As the data were not-normally distributed, outliers were removed by two different methods. This resulted in three versions of the data set: One containing all of the original data points; one with only data points inside the interquartile range (IQR), thus removing all data points outside the interval between the 25th and 75th percentile; and one with outliers removed using the median absolute deviation method (MAD) with the conservative threshold value of three [24]. All analyses were conducted using data in which all three observation sessions were pooled for each treatment and each individual separately. Prior to this it was investigated if the three observation sessions from the same treatment and individual originated to the same distribution. This analysis showed that for some behaviors that data did not belong to the same distribution and should therefore theoretically not be pooled. However, when comparing the results for the observation sessions separately the results were highly similar to the results found when pooling the data. We have therefore, chosen to only present the methods and results for the pooled data.

2.4. Proportion of Time Each Individual Spent on Each Behavior

The proportion of time each individual spent on each behavior was estimated for the different observation sessions in order to examine the differences in the distribution of time spent on each behavior, both between treatments and individuals. Furthermore, χ^2 tests with Yates corrections [25] were carried out on pooled data, with the variables being the different treatments and the two individuals (Appendix B). This was only carried out for the data set containing all data points, as it was only for this data set that all nine hours were represented.

2.5. Reaction Norms for Testing Differences between Individuals and Between Treatments

For all data sets, the medians, variances, asymmetry indices (skewness) and kurtoses were calculated to examine the differences in time each behavior lasted per occurrence, how much it varied and the shape of the data between individuals and treatments. Due to the non-normal distribution of the data, the variances were based on the IQR.

For each behavior, the medians for both individuals and treatments were plotted along with a trend line between the median of treatment C and median of treatment D for each individual. The slopes of the trend lines were calculated as well as the percentage differences in the trend line slopes between the two individuals for the same behavior. This procedure was also carried out for the variances, asymmetry indices and kurtoses of the pooled data for the data set containing all data points. The same plots were made for the two data sets where outliers had been removed (Appendix C). The slopes of the trend lines of these variables portray the two individuals' behavioral reaction norms i.e., the set of behavioral phenotypes that a single genotype produces in a given set of environments [6].

Furthermore, $\chi 2$ tests were carried out to compare all variables for both the individuals under the same treatment and the different treatments for each individual (Appendix D).

Finally, due to the short observation period, the randomized moving average of medians and variances were calculated and plotted in order to confirm the reliability of the results (Appendix E).

3. Results

3.1. Proportion of Time Each Individual Spent on Each Behavior

The time spent on different behaviors varied between all the observation sessions and the individuals. Figure 1 shows that individual 2 generally spent a greater amount of time on the behavior 'stereotypic' and a smaller amount of time on 'inactive' behavior compared to individual 1. However, the amount of time the two individuals spent on these behaviors varied greatly between observation sessions. When comparing the two individuals' 'stereotypic' and 'inactive' behavior for treatment D, a significant difference, between the two individuals, was observed for both behaviors. For this treatment, individual 1 spent a greater amount of time being 'inactive' than individual 2 ($p < 0.05$). The opposite was found for the amount of time the individuals spent on 'stereotypic' behavior, meaning that individual 2 spent more time on 'stereotypic' behavior than individual 1 ($p < 0.01$) (see Appendix B). Furthermore, it was found that individual 1 spent significantly more time being 'inactive' for treatment D in comparison to treatment C ($p < 0.05$) (see Appendix B).

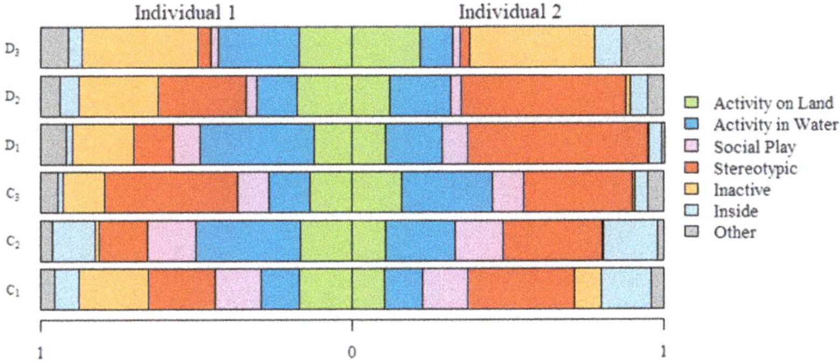

Figure 1. Proportion of time each individual spent on the different behaviors for each of the three observation sessions for each treatment (C = control, D = dog-scented object). The three lower bars represent the control observation sessions and the three upper bars represent the observation sessions in which the individuals were exposed to olfactory stimuli. The data were pooled and compared by χ^2 tests with a Yates correction (see Appendix B).

3.2. Reaction Norms for Testing Differences between Individuals and Treatments

An increase in the median time spent on each behavior between treatment C and D could be observed for both individuals and all behaviors, except the median time individual 2 spent 'inside', which showed a decrease between treatment C and D ($p < 0.01$) (Figure 2) (see Appendix D). For the three behaviors of 'activity in water', 'stereotypic' and 'inactive', significant differences in the median time were found between the two treatments for each individual ($p < 0.01$) (see Appendix D). When comparing the median time spent on 'stereotypic' behavior, a significant difference was found between the individuals for treatment D ($p < 0.001$) but not for treatment C (see Appendix D). For the behavior 'inactive', a significant difference was observed between the individuals for treatment C ($p < 0.001$). For most behaviors, it was found, for both individuals, that the variances increased between treatment C and D (Figure 2) (see Appendix D). The opposite was found for the behavior stereotypic' of individual 1 and the behavior 'inside' for individual 2 ($p < 0.01$), meaning that the variances decreased between treatment C and D for these combinations (see Appendix D). For the behaviors 'inactive' and 'inside', significant differences were found between the variances of both individuals ($p < 0.05$) and between those of the two treatments ($p < 0.01$) (see Appendix D). There were also significant differences found between the variances of time spent on 'stereotypic' behavior between the two individuals for both treatments ($p < 0.01$) and between the two treatments for individual 2 ($p < 0.001$) (see Appendix D).

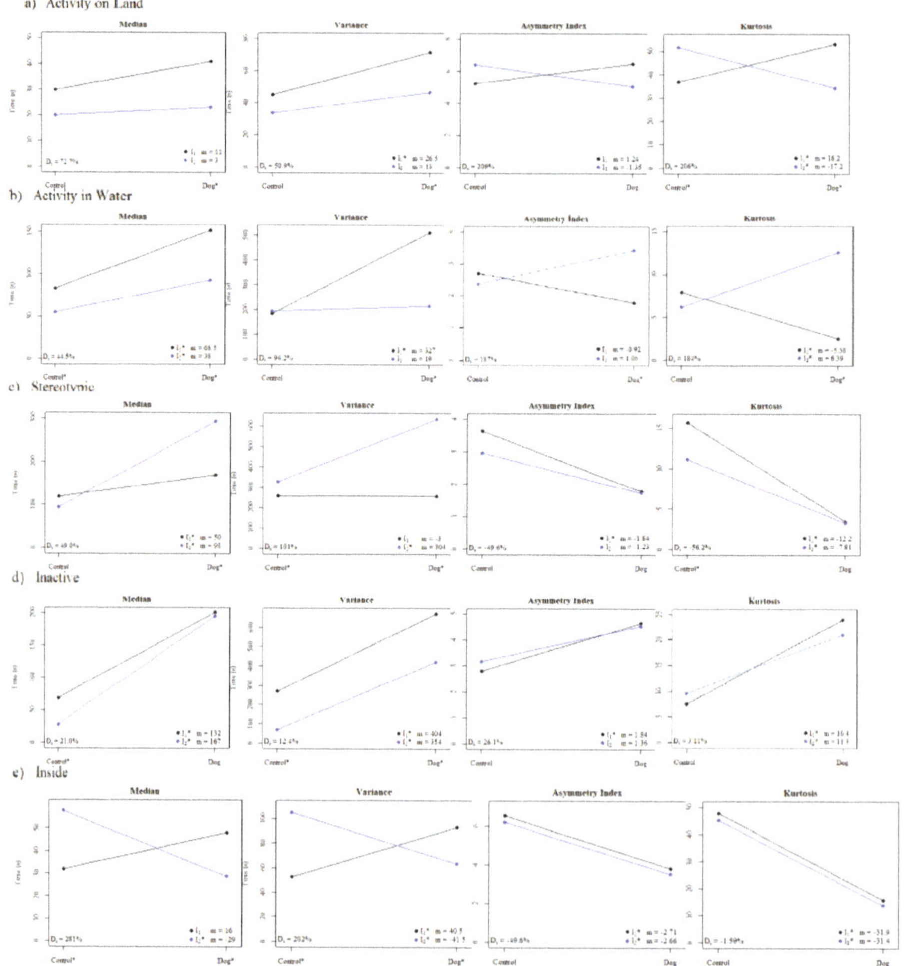

Figure 2. Behavioral reaction norms; for each individual, the median and variance of time spent on a given behavior are shown for treatment C and for treatment D along with trend lines between the medians and between the variances of the two treatments for the same individual. The asymmetry index and kurtosis are also shown for treatment C and for treatment D and each individual along with trend lines between the asymmetry indices and between the kurtoses of the two treatments for the same individual. The medians, variances, asymmetry indices and kurtoses are based on pooled data. The slope (m) and difference in slope in percent (D_S) are given for each comparison. The medians, variances, asymmetry indices and kurtoses were compared by χ^2 tests with a Yates correction. Comparisons in which the χ^2 test resulted in significant results between the two individuals for the same treatment are indicated by * next to the relative treatment. Comparisons in which the χ^2 test resulted in significant results between the two treatments for the same individual are indicated by * next to the relative individual. For further details on χ^2 values, see Appendix D.

When comparing the asymmetry indices of the two treatments, results varied greatly for both individuals in terms of the asymmetry index between the two treatments (Figure 2) (see Appendix D). Significant differences between the asymmetry indices of the two treatments were found for the behaviors 'stereotypic', 'inactive' and 'inside' of individual 1 ($p < 0.05$), whereas for individual 2 it was only the behavior 'inside' that showed a significant difference ($p < 0.01$). A significant difference between the individuals for treatment D for 'activity in water' was also found, where individual 2 had a significant higher asymmetry index than individual 1 ($p < 0.05$). Furthermore, no significant differences were found between the two individuals for either of the two treatments (see Appendix D).

Similar to the asymmetry indices, great variation in whether the slope was positive or negative when comparing the kurtoses of the two treatments was also found for the different behaviors (Figure 2) (see Appendix D). For all behaviors, significant differences were found between the kurtoses of the two treatments for both individuals ($p < 0.001$). When comparing the kurtoses of the two individuals for treatment C, significant differences were found for the behaviors 'activity on land' and 'stereotypic' ($p < 0.01$) (see Appendix D). For treatment D, significant differences were found between the kurtoses of the two individuals for the behaviors 'activity on land' and 'activity in water' ($p < 0.001$) (see Appendix D).

The randomized moving average of the medians show that the medians of each behavior stabilize within the three observation sessions for both individuals and both treatments. The same was found for the randomized moving average of the variances (see Appendix E).

4. Discussion

4.1. Results of the Case Study

The results of the case study demonstrate the value of behavioral instability as a new quantitative method of behavior assessment. In this case study, an increase in median time and variance was found for most behaviors when the individuals were exposed to the olfactory stimuli of dog odor. This indicates that the occurrences of a behavior generally lasted longer when the individuals were provided with the olfactory stimuli, but also that the individuals were less predictable during the time they were engaged in each occurrence of a behavior. The effect of stimuli on the asymmetry index and kurtosis varied greatly between the individuals and behaviors. This demonstrates that there was a variation in predictability for the behaviors of both individuals when exposed to the olfactory stimuli.

The difference found in the two individuals' responses to olfactory stimuli is a good example of how individuals can respond differently to environmental stress. This exhibits how the understanding of different behavioral reaction norms is important in the evaluation of welfare in captive animals [7,26], implying that different individuals can benefit from different types of enrichment in order to increase their welfare. When exposed to olfactory stimuli, there was a significant difference between the two individuals in the amount of time each spent on 'stereotypic' and 'inactive' behavior (Appendix B). One individual spent less time being 'stereotypic' and more time on 'inactive' behavior, while the other individual spent less time being 'inactive' and more time on 'stereotypic' behavior (Figure 1). The same was found when comparing the quantitative variables—median, variance, asymmetry index and kurtosis—of the data for the two individuals. This analysis showed significant differences in medians for treatment D and variances for treatment C and treatment D of time spent on 'stereotypic' behavior between the individuals. These differences were larger when the individuals were exposed to olfactory stimuli (Figure 2). This demonstrates that the individuals responded differently to the stimulus, supporting the statement that individuals with different behavioral reaction norms react differently to the same stimulus, as they often have different ways of coping with changes in their environment [9]. When comparing the asymmetry indices of both individuals for 'stereotypic' and 'inactive' behavior it was found that there was a smaller difference between the individuals, when exposed to stimuli, than under normal conditions; this means that the distributions were more similar. These various results indicate the importance of using different quantitative variables.

4.2. Reliability of Results

Despite the fact that this investigation has been conducted within a relatively short period of time, the number of seconds in which the two individuals were observed (194,400 seconds per individual) is large compared to other previous studies where the instantaneous sampling technique has been utilized; see for example [1] with six individuals and with 19,200 seconds of observation per individual; [10] with 55 individuals and 17,472 seconds of observation per individual; [12] with two individuals and 19,200 seconds of observation per individual. There is only one study where the number of seconds of observation was higher than in our investigation; [11] with 10,965,600 seconds of observation but conducted on a single individual and over a long period.

Furthermore, the randomized moving average of medians and variances also show that the medians and variances of each behavior stabilize within the three observation sessions of each treatment (Appendix E). Therefore, we believe that we have provided a robust preliminary dataset where the genetic and environmental bias are minimized, as the two individuals were sisters and the period of investigation is very short, therefore less prone to environmental fluctuations. All these factors allow us to draw robust conclusions. At the same time, we have provided a solid theoretical framework which can be applied to behavioral studies in the immediate future.

4.3. Considerations when Removing Outliers

The results discussed were generally observed for all three data sets, but some slight differences were found due to the removal of outliers. When using the MAD method, only large values were indicated as outliers and removed due to the distribution of the data, whereas when using the IQR to identify outliers, an equal amount of values smaller and greater than the median were removed. When removing outliers using IQR, only the most frequent results are shown; it can be argued that this gives a better representation of the data. A similar argument presents itself when removing outliers using MAD, as this method removes extreme values that have a small likelihood of occurring. When studying behavior, the distribution of the data is usually skewed to the right; hence, the removal of outliers using MAD can remove important information, as an individual performing a behavior for a long time is also a part of their behavior and cannot simply be ignored [27]. Even though removing outliers presents some disadvantages, it can also be a resourceful tool when comparing individuals and treatments, since the removal of outliers can increase the amount of significant results. However, the original data set should always be analyzed as well. Ideally, behavioral data should be analyzed both with and without outliers, as the different methods supplement each other.

4.4. Applying Behavioral Instability to Behavioral Investigations

In this study, it is shown how behavioral instability can be applied to behavioral observations by investigating the median, variance, asymmetry indices and kurtosis of different behaviors. The results of this study supports that behavioral instability can be applied to a more traditional type of behavioral data, in which an ethogram and observations of different behaviors are used. Thus, behavioral instability can be introduced as a new quantitative method for analyzing traditional ethograms. This study used this new method along with the traditional methods, enabling a comparison of the two methods. One of the major issues when using the traditional methods for studying behavior is the lack of comparable systematic and quantitative results [13,16]. Traditional methods are primarily used to estimate the percentage of time spent on various activities [17]. This estimate is highly dependent on the ethogram used, as the percentage of time spent on one activity is always dependent on the amount of time spent on other activities. Comparisons between studies are, therefore, only possible if highly similar ethograms are used, which can prove difficult in the comparison of behavior between species. The application of the concept behavioral instability enables the comparison of behavior regardless of differences in ethograms. This is possible due to the method's quantitative approach that uses the median, variance, asymmetry index and kurtosis. The advantage of this approach is that these variables for one behavior are less dependent on the other behaviors.

The traditional methods also lack a protocol ensuring systematic data sampling. The results of this study indicate the need for longer observation sessions, as short observation sessions lead to a higher risk of type II errors. However, if the data are symmetric-leptokurtic, the risk of type II errors is lower, thus making it possible to create a behavioral analysis based on short observation sessions. When applying the concept of behavioral instability to behavioral studies, the data should be sampled using continuous focal sampling over the entire day since the occurrence of different behaviors has been observed to change throughout the day. The results of this study showed that many behaviors occurred for both shorter and longer periods of time and, therefore, information can be lost when using sampling techniques such as instantaneous sampling. Altmann [17] states that instantaneous sampling is primarily used for studying the proportion of time spent on various activities. However, the results would not be accurate, as behaviors shorter than the time between two preselected sampling instances would most likely not be recorded. When using this new quantitative method of applying behavioral instability, it is, therefore, important that sampling is conducted throughout the entire day using continuous focal sampling.

While the quantitative results of this new method enable comparisons between studies, traditional methods should not be dismissed, as valuable information also lies in knowing when an individual performs various behaviors throughout the day and the proportion of the day spent on different behaviors. It is suggested that the two methods are used collaboratively, comparing and combining the results of both approaches, in order to obtain the most reliable results. The application of the concept behavioral instability to traditional behavioral analyses allows quantitative data collection. This can provide researchers with a relatively unbiased evaluation of behavioral responses and the effectiveness of enrichment manipulation, which can contribute to the improvement of enrichment programs and animal welfare in captivity [13]. It has been debated whether the study of behavioral reaction norms can provide new insights for the field of behavioral ecology [7]. The use of behavioral instability as a new quantitative and systematic method for studying behavioral responses could be highly relevant when studying animal conservation. When captive populations are being managed with the aim of re-introducing individuals to the wild, an understanding of the behavioral reaction norms can provide insight on how to conserve behavioral responses that could be beneficial in the wild.

Author Contributions: All authors have contributed equally and have conceptualized the project and done the methodology. C.P. and T.H.J. provided the resources, while A.C.L., A.G., H.L. and M.G.L. did the investigation and data curation. A.C.L., A.G. and H.L. did the formal analysis and visualization and took the lead in writing the manuscript. All authors provided critical feedback and helped shape the research, analysis and manuscript. C.P. and T.H.J. were responsible for the project. All authors have read and agreed to the published version of the manuscript.

Funding: This investigation was supported by the Aalborg Zoo Conservation Foundation (Grant Number: 3-2018).

Acknowledgments: The authors wish to thank the staff of Aalborg Zoo, with special thanks to Katrine Christensen and Frank Thomsen for their time and involvement in this study. We also thank three anonymous reviewers and Rabby Zhang for invaluable suggestions and help.

Conflicts of Interest: The authors declare no conflict of interest.

Appendix A

Map of the polar bear enclosure at Aalborg Zoo. Only one exhibit was used in the study, this exhibit is indicated on the map. The other exhibit housed the mother of the two polar bears studied. The placement of the four cameras used for surveillance is shown.

Appendix B

χ^2 Test of Total Time

χ^2 test results for testing the percentages of time spent on each behavior for the entire observation session based on pooled data for treatment C and treatment D of each individual. A blank space indicates that the behavior consisted of less than 5% of the observation session. Results when comparing individual 1 with individual 2 are shown on the left and results comparing treatment C and D are shown on the right.

Behavior	χ^2 Test Comparing Individuals		χ^2 Test Comparing Treatments	
	C_{pooled}	D_{pooled}	$I_{1_{pooled}}$	$I_{2_{pooled}}$
Activity on land	$\chi^2 = 0.3542$ $p = 0.5518$	$\chi^2 = 6.730 \times 10^{-3}$ $p = 0.9346$	$\chi^2 = 1.388 \times 10^{-4}$ $p = 0.9906$	$\chi^2 = 0.2510$ $p = 0.6164$
Activity in water	$\chi^2 = 0.05417$ $p = 0.8160$	$\chi^2 = 2.058$ $p = 0.1514$	$\chi^2 = 0.6791$ $p = 0.4099$	$\chi^2 = 0.7223$ $p = 0.3954$
Social play	$\chi^2 = 2.666 \times 10^{-4}$ $p = 0.9870$			
Stereotypic	$\chi^2 = 0.8345$ $p = 0.3610$	$\chi^2 = 9.695$ $p = 0.001848$	$\chi^2 = 3.104$ $p = 0.07808$	$\chi^2 = 0.2462$ $p = 0.6197$
Inactive		$\chi^2 = 4.501$ $p = 0.03388$	$\chi^2 = 5.736$ $p = 0.01661$	
Inside	$\chi^2 = 1.118$ $p = 0.2904$			$\chi^2 = 2.169$ $p = 0.1408$

Appendix C

Appendix C.1. Slopes of Medians, Variances, Asymmetry Indices and Kurtoses

For each individual the median time and variance of time spent on a given behavior are shown for treatment C and for treatment D along with trend lines between the medians and the variances for the two individuals. The asymmetry index and kurtosis are also shown for treatment C and for treatment D and each individual along with trend lines between the asymmetry indices and kurtoses of the two treatments for the same individual. The medians, variances, asymmetry indices and kurtoses are based in pooled data. The slope (m) and difference in slope in percent (D_s) are given for each comparison. The medians, variances, asymmetry indices and kurtoses were compared by χ^2 test with Yates correction. Comparisons in which the χ^2 test resulted in significant results between the two individuals for the same treatment are indicated by * next to the relative treatment. Comparisons in which the χ^2 test resulted in significant results between the two treatments for the same individual are indicated by * next to the relative individual. The medians are not shown for the dataset where outliers were removed outside IQR since the values are identical to the values shown in figures for the data set with all data points.

Appendix C.2. Dataset where Outliers were Removed outside IQR

Activity on land

Activity in water

Stereotypic

Inactive

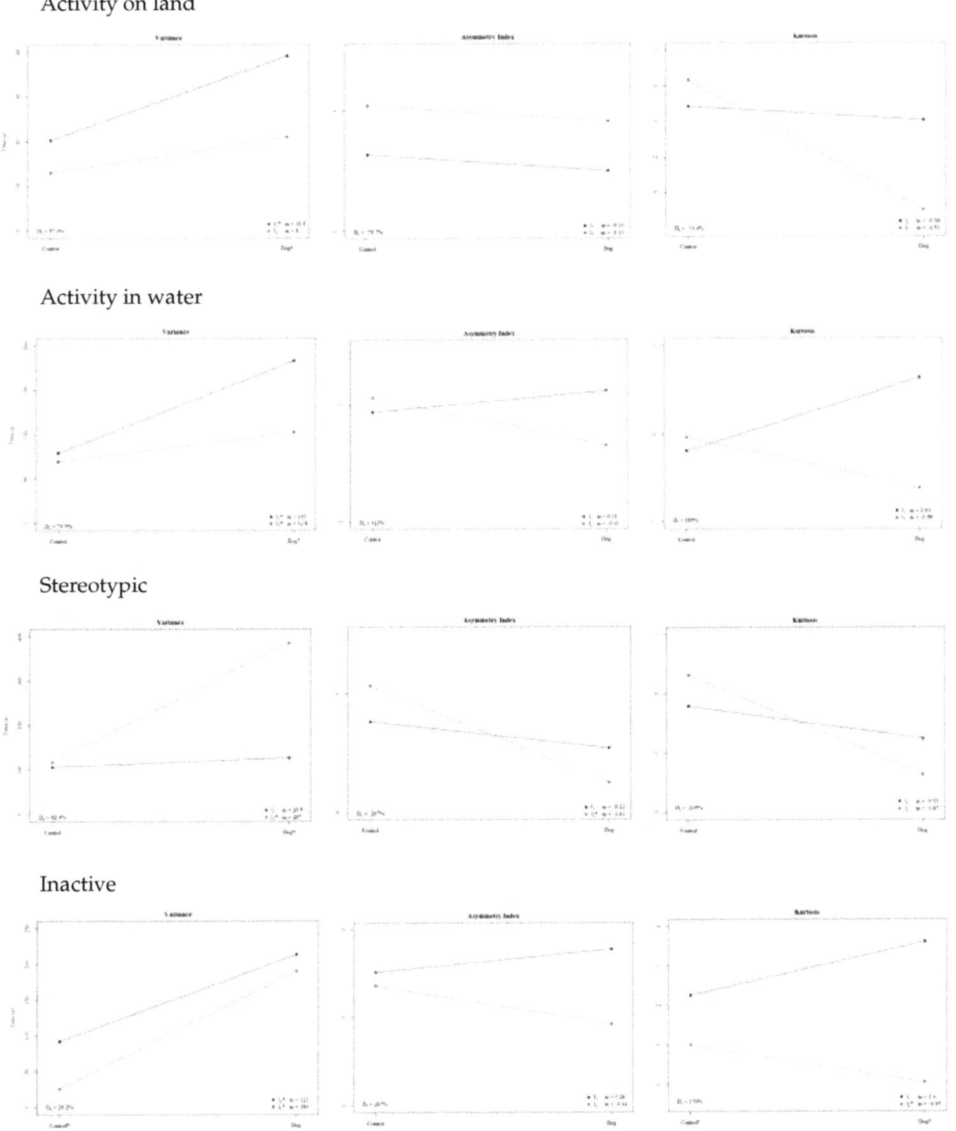

Inside

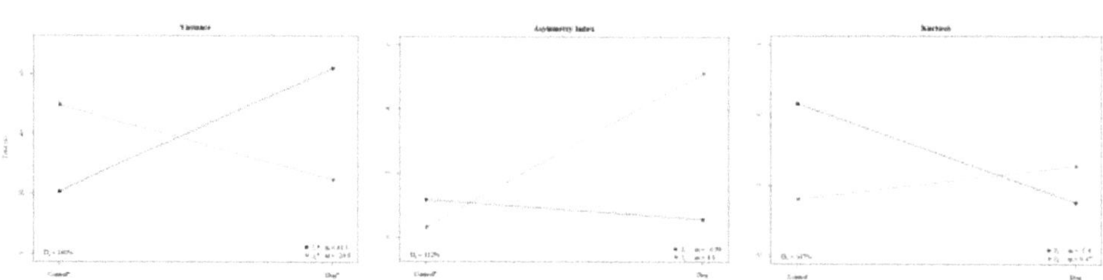

Appendix C.3. Dataset where Outliers were Removed Using the MAD Method

Activity on land

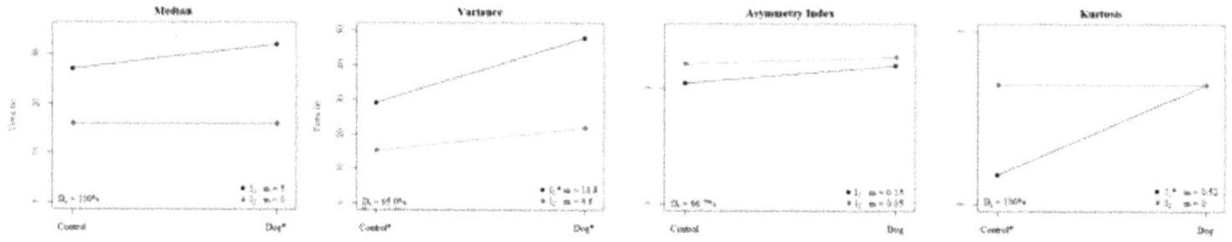

Activity in water

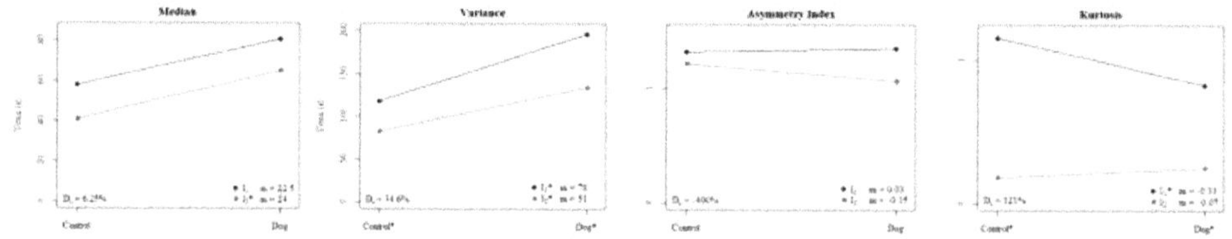

Stereotypic

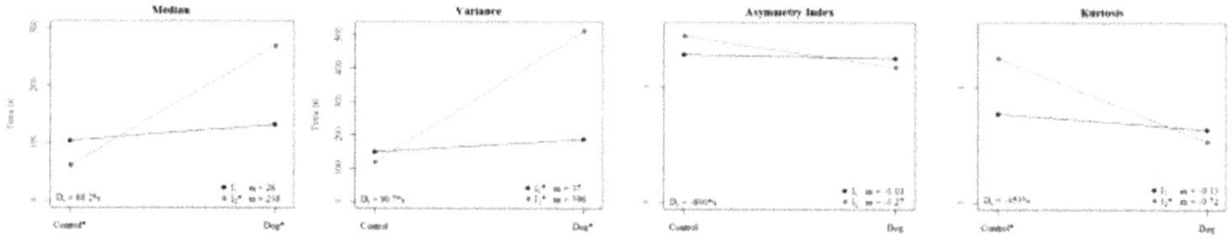

Inactive

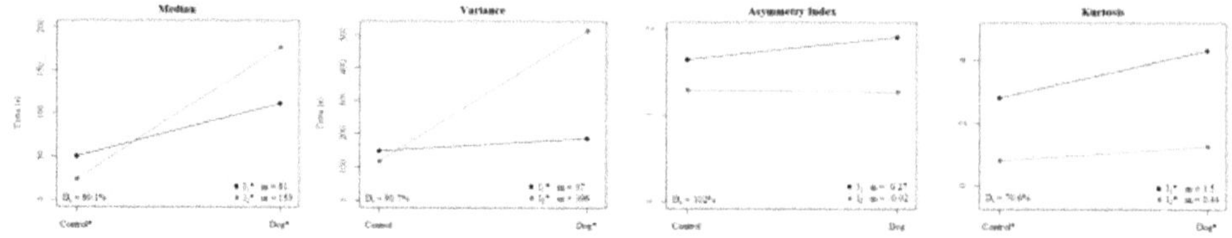

Inside

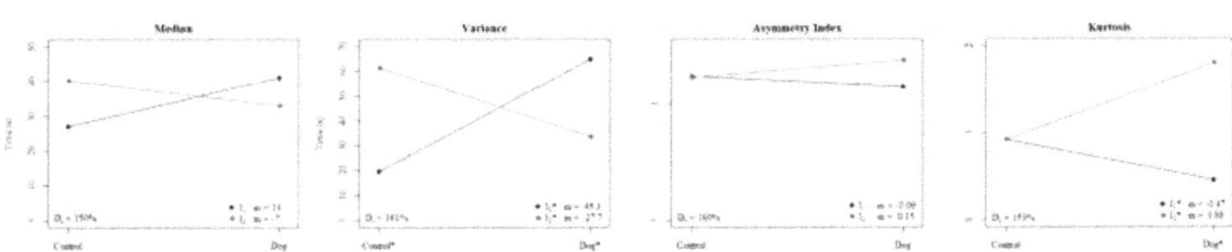

Appendix D

χ^2 Test for Distribution of Statistics

χ^2 test results for comparing the medians, variances, asymmetry indices and kurtoses of the pooled data for each behavior for the different data sets. The pooled data for asymmetry indices and kurtoses were multiplied by 10 before the perpetration of the χ^2 test. Results when comparing individual 1 and individual 2 are shown on the left and results comparing treatment C and D are shown on the right. Comparisons for which χ^2 test could not be performed due to negative values are indicated by -. χ^2 tests for medians for the data set where outliers were removed outside IQR are not shown since the medians are the same as for the dataset containing all data points.

D.1.a: χ^2 Tes t of medians, all data points

Behavior	χ^2 Test Comparing Individuals		χ^2 Test Comparing Treatments	
	C_{pooled}	D_{pooled}	$I1_{pooled}$	$I2_{pooled}$
Activity on land	$\chi^2 = 2.000$ $p = 0.1573$	$\chi^2 = 5.063$ $p = 0.02445$	$\chi^2 = 1.704$ $p = 0.1917$	$\chi^2 = 0.2093$ $p = 0.6473$
Activity in water	$\chi^2 = 5.681$ $p = 0.01715$	$\chi^2 = 14.00$ $p = 0.0001831$	$\chi^2 = 20.01$ $p = 7.705 \times 10^{-6}$	$\chi^2 = 9.757$ $p = 0.001787$
Stereotypic	$\chi^2 = 2.692$ $p = 0.1009$	$\chi^2 = 33.28$ $p = 7.974 \times 10^{-9}$	$\chi^2 = 8.681$ $p = 0.003216$	$\chi^2 = 101.0$ $p < 2.2 \times 10^{-16}$
Inactive	$\chi^2 = 17.33$ $p = 3.142 \times 10^{-5}$	$\chi^2 = 0.090909$ $p = 0.7630$	$\chi^2 = 64.53$ $p = 9.491 \times 10^{-16}$	$\chi^2 = 125.1$ $p < 2.2 \times 10^{-16}$
Inside	$\chi^2 = 7.511$ $p = 0.006130$	$\chi^2 = 4.6883$ $p = 0.03037$	$\chi^2 = 3.200$ $p = 0.07364$	$\chi^2 = 9.667$ $p = 0.001876$

D.1.b: χ^2 test of variances, all data points

Behavior	χ^2 Test Comparing Individuals		χ^2 Test Comparing Treatments	
	C_{pooled}	D_{pooled}	$I1_{pooled}$	$I2_{pooled}$
Activity on land	$\chi^2 = 1.607$ $p = 0.2049$	$\chi^2 = 5.1802$ $p = 0.02285$	$\chi^2 = 6.0279$ $p = 0.01408$	$\chi^2 = 2.0994$ $p = 0.1474$
Activity in water	$\chi^2 = 0.3485$ $p = 0.5550$	$\chi^2 = 120.68$ $p < 2.2 \times 10^{-16}$	$\chi^2 = 153.85$ $p < 2.2 \times 10^{-16}$	$\chi^2 = 0.92631$ $p = 0.3358$
Stereotypic	$\chi^2 = 8.000$ $p = 0.004677$	$\chi^2 = 160.06$ $p < 2.2 \times 10^{-16}$	$\chi^2 = 0.02381$ $p = 0.8774$	$\chi^2 = 96.901$ $p < 2.2 \times 10^{-16}$
Inactive	$\chi^2 = 117.65$ $p < 2.2 \times 10^{-16}$	$\chi^2 = 56.668$ $p = 5.15910^{-14}$	$\chi^2 = 172.9$ $p < 2.2 \times 10^{-16}$	$\chi^2 = 254.14$ $p < 2.2 \times 10^{-16}$
Inside	$\chi^2 = 17.5$ $p = 2.873 \times 10^{-5}$	$\chi^2 = 5.5607$ $p = 0.01837$	$\chi^2 = 11.273$ $p = 0.0007863$	$\chi^2 = 10.221$ $p = 0.001388$

D.1.c: χ^2 test of asymmetry indices, all data points

Behavior	χ^2 Test Comparing Individuals		χ^2 Test Comparing Treatments	
	$\mathbf{C}_{\text{pooled}}$	$\mathbf{D}_{\text{pooled}}$	$\mathbf{I1}_{\text{pooled}}$	$\mathbf{I2}_{\text{pooled}}$
Activity on land	$\chi^2 = 1.201$ $p = 0.2731$	$\chi^2 = 1.713$ $p = 0.1905$	$\chi^2 = 1.311$ $p = 0.2523$	$\chi^2 = 1.588$ $p = 0.2076$
Activity in water	$\chi^2 = 0.2064$ $p = 0.6496$	$\chi^2 = 5.254$ $p = 0.0219$	$\chi^2 = 1.886$ $p = 0.1697$	$\chi^2 = 1.934$ $p = 0.1644$
Stereotypic	$\chi^2 = 0.6759$ $p = 0.411$	$\chi^2 = 0.007748$ $p = 0.9299$	$\chi^2 = 6.280$ $p = 0.01221$	$\chi^2 = 3.216$ $p = 0.07293$
Inactive	$\chi^2 = 0.2501$ $p = 0.617$	$\chi^2 = 0.01137$ $p = 0.9151$	$\chi^2 = 4.58$ $p = 0.03235$	$\chi^2 = 2.384$ $p = 0.1226$
Inside	$\chi^2 = 0.07902$ $p = 0.7786$	$\chi^2 = 0.09389$ $p = 0.7593$	$\chi^2 = 7.104$ $p = 0.007693$	$\chi^2 = 7.228$ $p = 0.007178$

D.1.d: χ^2 test of kurtoses, all data points

Behavior	χ^2 Test Comparing Individuals		χ^2 Test Comparing Treatments	
	$\mathbf{C}_{\text{pooled}}$	$\mathbf{D}_{\text{pooled}}$	$\mathbf{I1}_{\text{pooled}}$	$\mathbf{I2}_{\text{pooled}}$
Activity on land	$\chi^2 = 25.15$ $p = 5.305 \times 10^{-7}$	$\chi^2 = 39.02$ $p = 4.187 \times 10^{-10}$	$\chi^2 = 29.34$ $p = 6.09 \times 10^{-8}$	$\chi^2 = 34.19$ $p = 4.996 \times 10^{-9}$
Activity in water	$\chi^2 = 2.062$ $p = 0.1511$	$\chi^2 = 66.84$ $p = 2.945\text{e-}16$	$\chi^2 = 27.65$ $p = 1.452 \times 10^{-7}$	$\chi^2 = 21.70$ $p = 3.196 \times 10^{-6}$
Stereotypic	$\chi^2 = 7.574$ $p = 0.005922$	$\chi^2 = 0.02510$ $p = 0.8741$	$\chi^2 = 77.88$ $p < 2.2 \times 10^{-16}$	$\chi^2 = 42.31$ $p = 7.788 \times 10^{-11}$
Inactive	$\chi^2 = 2.824$ $p = 0.09284$	$\chi^2 = 1.915$ $p = 0.1664$	$\chi^2 = 86.12$ $p < 2.2 \times 10^{-16}$	$\chi^2 = 41.71$ $p = 1.059 \times 10^{-10}$
Inside	$\chi^2 = 0.6862$ $p = 0.4075$	$\chi^2 = 1.432$ $p = 0.2315$	$\chi^2 = 159.7$ $p < 2.2 \times 10^{-16}$	$\chi^2 = 167.1$ $p < 2.2 \times 10^{-16}$

D.2.b: χ^2 test of variances, IQR

Behavior	χ^2 Test Comparing Individuals		χ^2 Test Comparing Treatments	
	$\mathbf{C}_{\text{pooled}}$	$\mathbf{D}_{\text{pooled}}$	$\mathbf{I1}_{\text{pooled}}$	$\mathbf{I2}_{\text{pooled}}$
Activity on land	$\chi^2 = 1.581$ $p = 0.2086$	$\chi^2 = 5.400$ $p = 0.02014$	$\chi^2 = 5.924$ $p = 0.01486$	$\chi^2 = 1.882$ $p = 0.1701$
Activity in water	$\chi^2 = 0.5436$ $p = 0.4609$	$\chi^2 = 22.42$ $p = 2.194 \times 10^{-6}$	$\chi^2 = 41.12$ $p = 1.429 \times 10^{-10}$	$\chi^2 = 6.208$ $p = 0.0127$
Stereotypic	$\chi^2 = 0.6429$ $p = 0.4227$	$\chi^2 = 130.9$ $p < 2.2 \times 10^{-16}$	$\chi^2 = 1.766$ $p = 0.1839$	$\chi^2 = 141.3$ $p < 2.2 \times 10^{-16}$
Inactive	$\chi^2 = 36.92$ $p = 1.234 \times 10^{-9}$	$\chi^2 = 1.313$ $p = 0.2519$	$\chi^2 = 48.00$ $p = 4.255 \times 10^{-12}$	$\chi^2 = 124.5$ $p < 2.2 \times 10^{-16}$
Inside	$\chi^2 = 11.77$ $p = 0.0006032$	$\chi^2 = 15.99$ $p = 6.351 \times 10^{-5}$	$\chi^2 = 20.56$ $p = 5.771 \times 10^{-6}$	$\chi^2 = 8.25$ $p = 0.004075$

D.2.c: χ^2 test of asymmetry indices, IQR

Behavior	χ^2 Test Comparing Individuals		χ^2 Test Comparing Treatments	
	C_{pooled}	D_{pooled}	$I1_{pooled}$	$I2_{pooled}$
Activity on land	$\chi^2 = 0.9885$ $p = 0.3201$	$\chi^2 = 1.231$ $p = 0.2673$	$\chi^2 = 0.1648$ $p = 0.6848$	$\chi^2 = 0.0821$ $p = 0.7745$
Activity in water	$\chi^2 = 0.07778$ $p = 0.7803$	$\chi^2 = 1.211$ $p = 0.2711$	$\chi^2 = 0.1620$ $p = 0.6873$	$\chi^2 = 0.9592$ $p = 0.3274$
Stereotypic	$\chi^2 = 0.518$ $p = 0.4715$	$\chi^2 = 1.044$ $p = 0.3069$	$\chi^2 = 0.3828$ $p = 0.5361$	$\chi^2 = 5.069$ $p = 0.02436$
Inactive	$\chi^2 = 0.0781$ $p = 0.7799$	$\chi^2 = 2.675$ $p = 0.1019$	$\chi^2 = 0.2099$ $p = 0.6468$	$\chi^2 = 0.8363$ $p = 0.3605$
Inside	$\chi^2 = 4.791$ $p = 0.02861$	$\chi^2 = 0.04612$ $p = 0.8300$	$\chi^2 = 2.024$ $p = 0.1548$	$\chi^2 = 0.3942$ $p = 0.5300$

D.2.d: χ^2 test of kurtoses, IQR

Behavior	χ^2 Test Comparing Individuals		χ^2 Test Comparing Treatments	
	C_{pooled}	D_{pooled}	$I1_{pooled}$	$I2_{pooled}$
Activity on land	-	-	-	-
Activity in water	-	-	-	-
Stereotypic	-	-	-	-
Inactive	$\chi^2 = 4.858$ $p = 0.02752$	$\chi^2 = 34.39$ $p = 4.518 \times 10^{-9}$	$\chi^2 = 3.096$ $p = 0.07851$	$\chi^2 = 8.407$ $p = 0.003738$
Inside	-	-	-	-

D.3.a: χ^2 test of medians, MAD

Behavior	χ^2 Test Comparing Individuals		χ^2 Test Comparing Treatments	
	C_{pooled}	D_{pooled}	$I1_{pooled}$	$I2_{pooled}$
Activity on land	$\chi^2 = 2.814$ $p = 0.09345$	$\chi^2 = 5.333$ $p = 0.02092$	$\chi^2 = 0.4237$ $p = 0.5151$	$\chi^2 = 0$ $p = 1$
Activity in water	$\chi^2 = 2.919$ $p = 0.08753$	$\chi^2 = 1.651$ $p = 0.1988$	$\chi^2 = 3.6552$ $p = 0.05589$	$\chi^2 = 5.434$ $p = 0.01975$
Stereotypic	$\chi^2 = 11.21$ $p = 0.0008153$	$\chi^2 = 46.41$ $p = 9.609 \times 10^{-12}$	$\chi^2 = 3.435$ $p = 0.06385$	$\chi^2 = 131.1$ $p < 2.2 \times 10^{-16}$
Inactive	$\chi^2 = 9.1351$ $p = 0.0002507$	$\chi^2 = 15.125$ $p = 0.0001469$	$\chi^2 = 23.11$ $p = 1.528 \times 10^{-6}$	$\chi^2 = 116.5$ $p < 2.2 \times 10^{-16}$
Inside	$\chi^2 = 2.522$ $p = 0.1122$	$\chi^2 = 0.8649$ $p = 0.3524$	$\chi^2 = 2.882$ $p = 0.08956$	$\chi^2 = 0.6712$ $p = 0.4126$

D.3.b: χ^2 test of variances, MAD

Behavior	χ^2 Test Comparing Individuals		χ^2 Test Comparing Treatments	
	C_{pooled}	D_{pooled}	$I1_{pooled}$	$I2_{pooled}$
Activity on land	$\chi^2 = 4.373$ $p = 0.03873$	$\chi^2 = 9.727$ $p = 0.001816$	$\chi^2 = 4.580$ $p = 0.03234$	$\chi^2 = 1.142$ $p = 0.2853$
Activity in water	$\chi^2 = 6.0945$ $p = 0.01356$	$\chi^2 = 11.48$ $p = 0.000704$	$\chi^2 = 19.16$ $p = 1.203 \times 10^{-5}$	$\chi^2 = 11.99$ $p = 0.000536$
Stereotypic	$\chi^2 = 3.647$ $p = 0.05616$	$\chi^2 = 152.4$ $p < 2.2 \times 10^{-16}$	$\chi^2 = 4.179$ $p = 0.04093$	$\chi^2 = 247.5$ $p < 2.2 \times 10^{-16}$
Inactive	$\chi^2 = 22.38$ $p = 2.36 \times 10^{-6}$	$\chi^2 = 59.73$ $p = 1.091 \times 10^{-14}$	$\chi^2 = 51.22$ $p = 8.25 \times 10^{-13}$	$\chi^2 = 298.2$ $p < 2.2 \times 10^{-16}$
Inside	$\chi^2 = 21.59$ $p = 3.383 \times 10^{-6}$	$\chi^2 = 9.940$ $p = 0.001618$	$\chi^2 = 24.30$ $p = 8.229 \times 10^{-7}$	$\chi^2 = 8.127$ $p = 0.00436$

D.3.c: χ^2 test of asymmetry indices, MAD

Behavior	χ^2 Test Comparing Individuals		χ^2 Test Comparing Treatments	
	C_{pooled}	D_{pooled}	$I1_{pooled}$	$I2_{pooled}$
Activity on land	$\chi^2 = 0.1273$ $p = 0.7212$	$\chi^2 = 0.01984$ $p = 0.888$	$\chi^2 = 0.1$ $p = 0.7518$	$\chi^2 = 0.01004$ $p = 0.9202$
Activity in water	$\chi^2 = 0.03937$ $p = 0.8427$	$\chi^2 = 0.3240$ $p = 0.5692$	$\chi^2 = 0.003371$ $p = 0.9537$	$\chi^2 = 0.09825$ $p = 0.7539$
Stereotypic	$\chi^2 = 0.09412$ $p = 0.759$	$\chi^2 = 0.02645$ $p = 0.8708$	$\chi^2 = 0.003557$ $p = 0.9524$	$\chi^2 = 0.2793$ $p = 0.5972$
Inactive	$\chi^2 = 0.4181$ $p = 0.5179$	$\chi^2 = 1.288$ $p = 0.2564$	$\chi^2 = 0.2054$ $p = 0.6504$	$\chi^2 = 0.001563$ $p = 0.9685$
Inside	$\chi^2 = 0.0004049$ $p = 0.9839$	$\chi^2 = 0.2091$ $p = 0.6475$	$\chi^2 = 0.03389$ $p = 0.8539$	$\chi^2 = 0.08620$ $p = 0.7691$

D.3.d: χ^2 test of kurtoses, MAD

Behavior	χ^2 Test Comparing Individuals		χ^2 Test Comparing Treatments	
	C_{pooled}	D_{pooled}	$I1_{pooled}$	$I2_{pooled}$
Activity on land	$\chi^2 = 31.44$ $p = 2.055 \times 10^{-8}$	$\chi^2 = 0$ $p = 1$	$\chi^2 = 31.44$ $p = 2.055 \times 10^{-8}$	$\chi^2 = 0$ $p = 1$
Activity in water	$\chi^2 = 71.67$ $p < 2.2 \times 10^{-16}$	$\chi^2 = 31.14$ $p = 2.391 \times 10^{-8}$	$\chi^2 = 5.4724$ $p = 0.01932$	$\chi^2 = 1.140$ $p = 0.2858$
Stereotypic	$\chi^2 = 11.406$ $p = 0.0007321$	$\chi^2 = 1.034$ $p = 0.3092$	$\chi^2 = 1.199$ $p = 0.2736$	$\chi^2 = 29.12$ $p = 6.79 \times 10^{-8}$
Inactive	$\chi^2 = 108.5$ $p < 2.2 \times 10^{-16}$	$\chi^2 = 166.4$ $p < 2.2 \times 10^{-16}$	$\chi^2 = 31.51$ $p = 1.981 \times 10^{-8}$	$\chi^2 = 9.219$ $p = 0.002395$
Inside	$\chi^2 = 0$ $p = 1$	$\chi^2 = 81$ $p < 2.2 \times 10^{-16}$	$\chi^2 = 16.12$ $p = 5.932 \times 10^{-5}$	$\chi^2 = 28.47$ $p = 9.513 \times 10^{-8}$

Moving medians and variances

Plots showing the moving medians and variances of all the data points for each behavior. The moving medians and variances were calculated based on the pooled data for treatment C (blue) and treatment D (orange). The pooled data were randomized prior to calculating the moving medians and variances.

Appendix E

Appendix E.1. Moving Medians

Activity on land

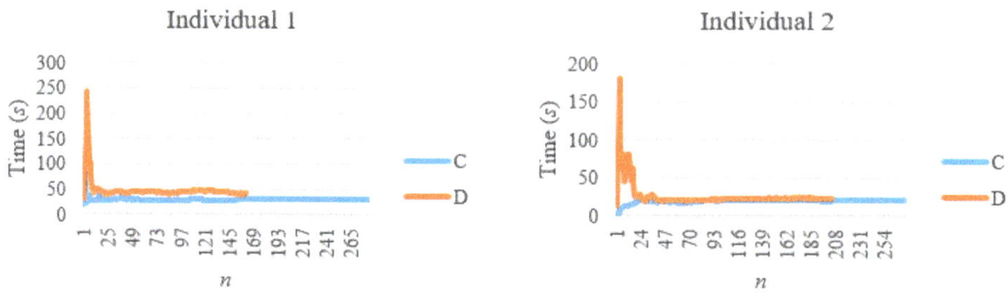

Activity in water

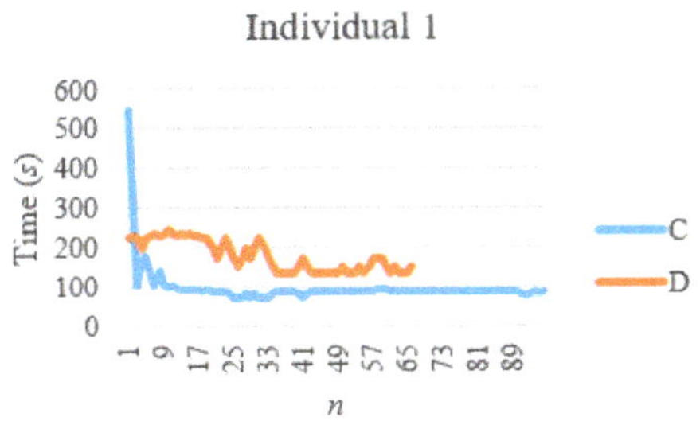

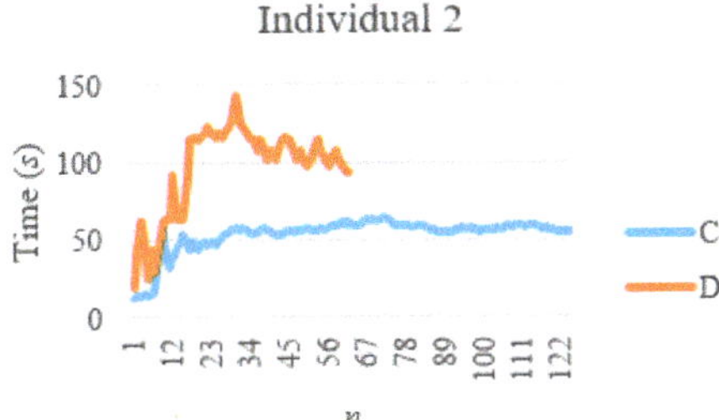

Stereotypic

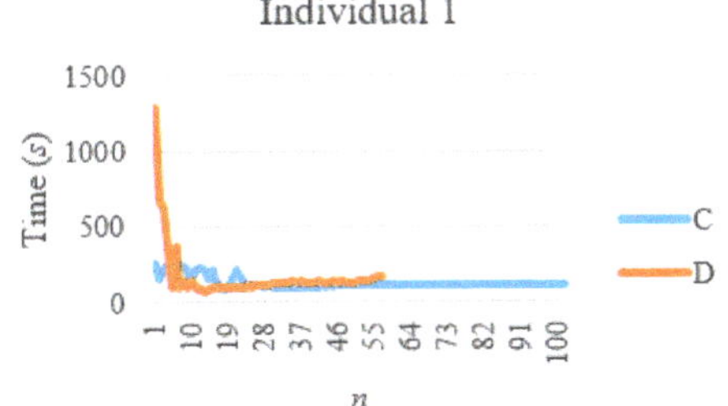

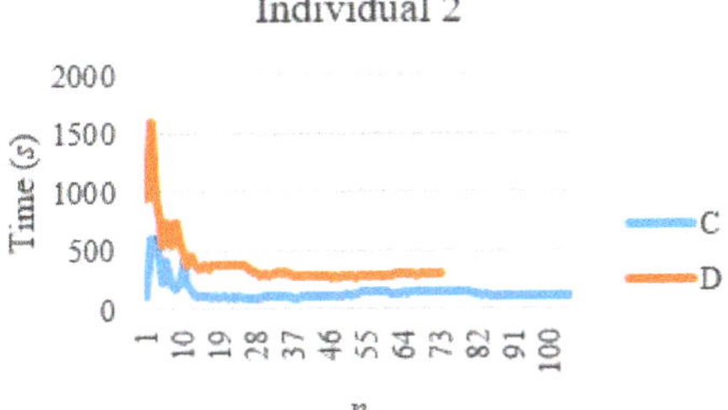

Social Play

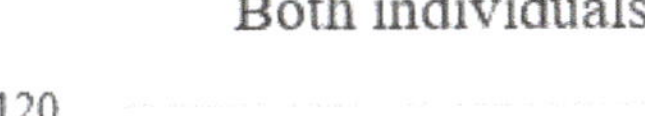

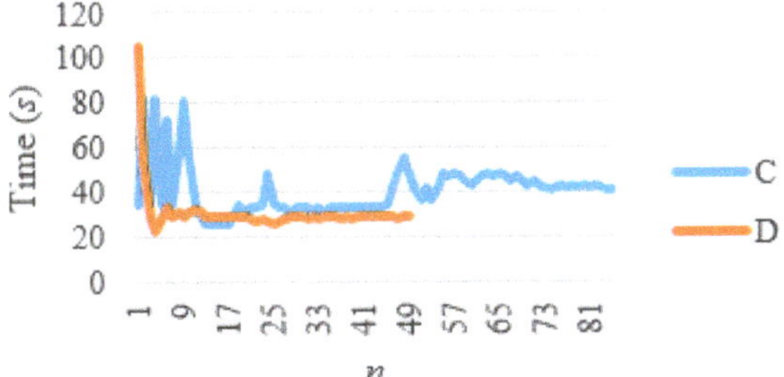

Inactive

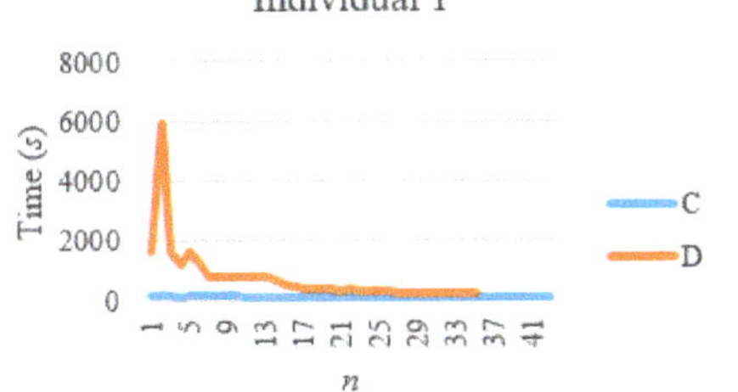

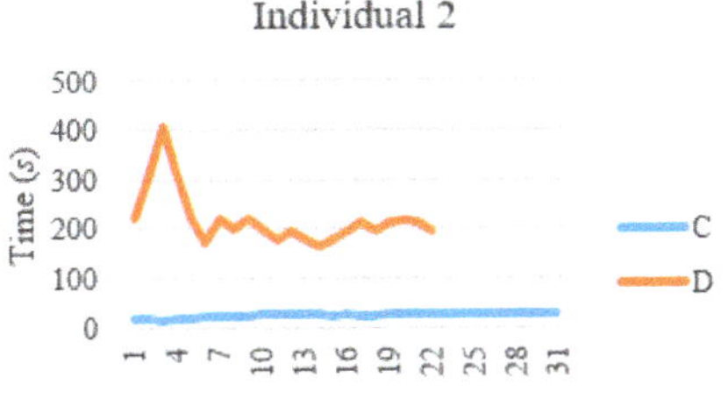

Inside

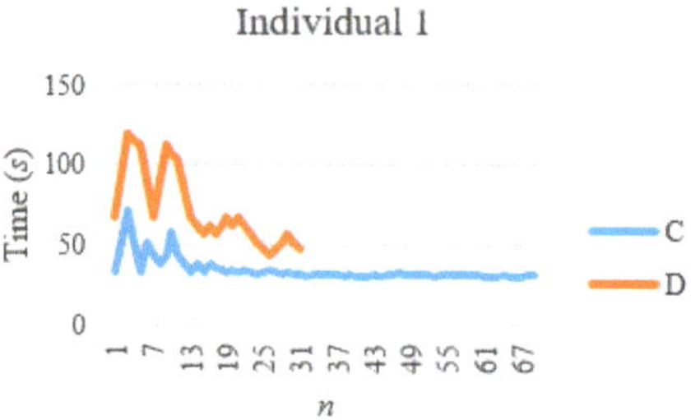

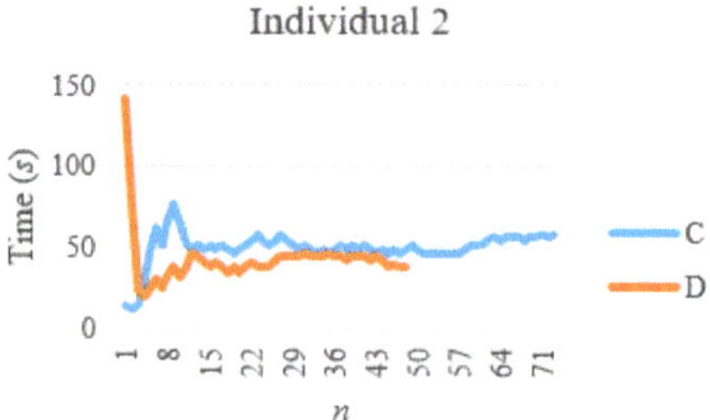

Appendix E.2. Moving Variances

Activity on land

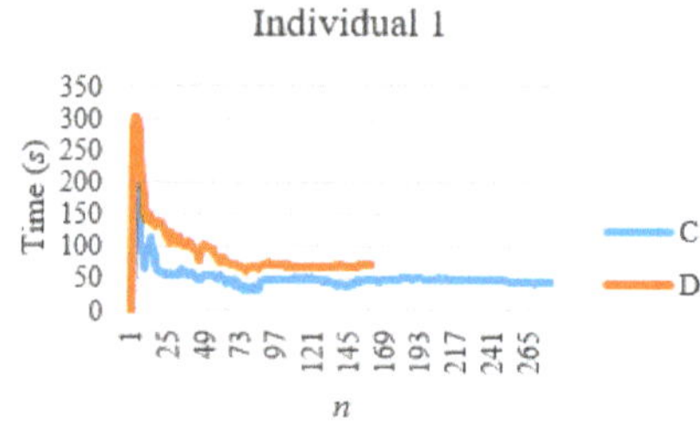

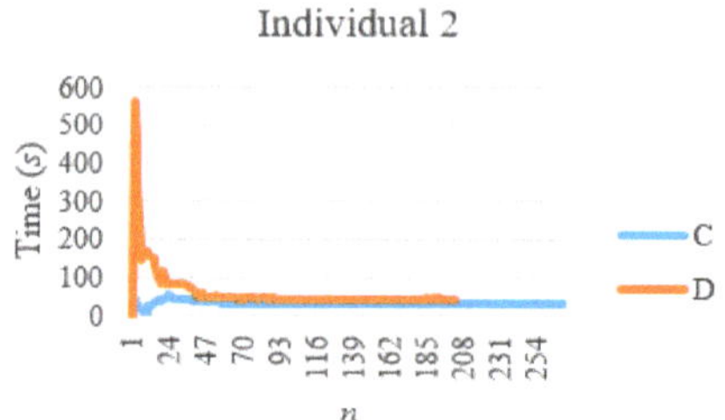

Activity in water

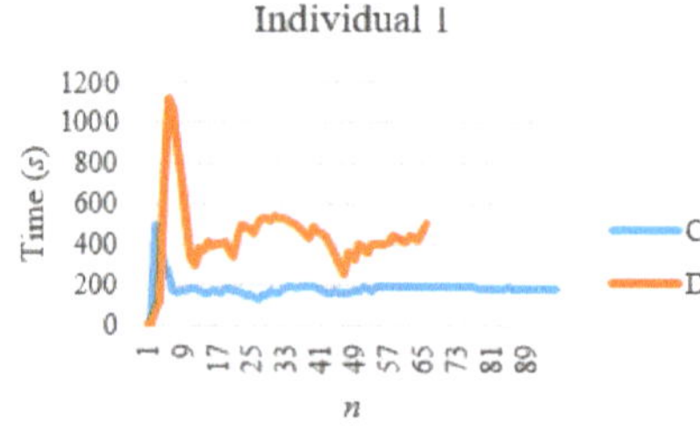

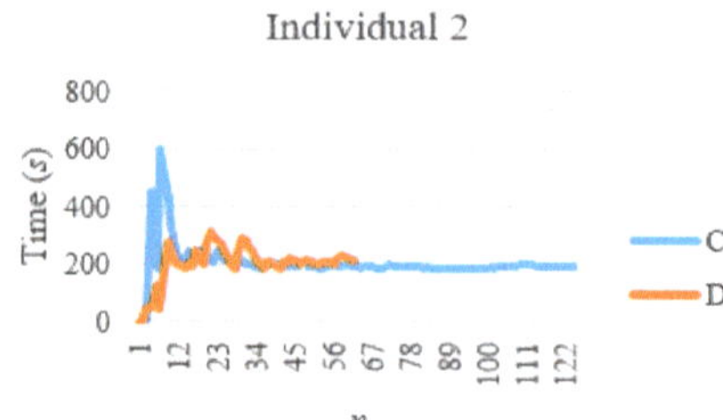

Stereotypic

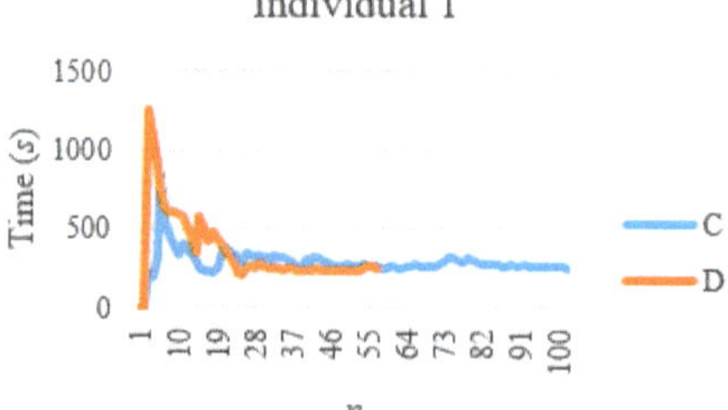

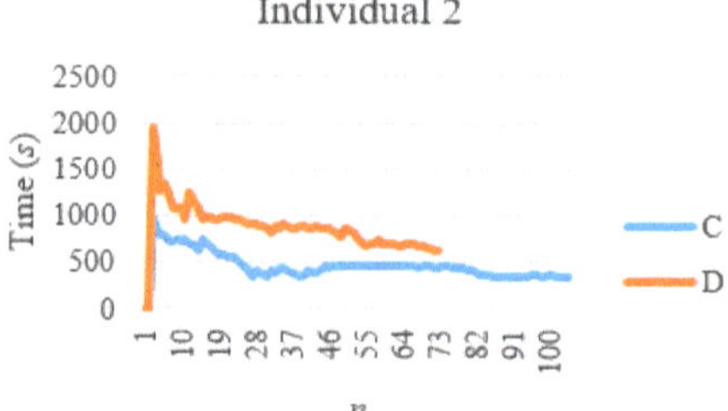

Social Play

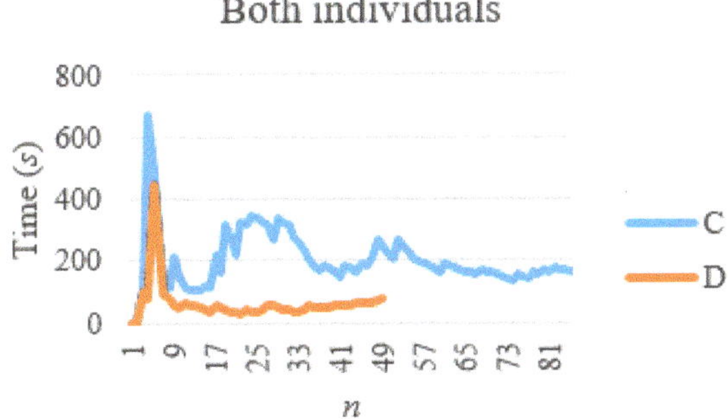

Inactive

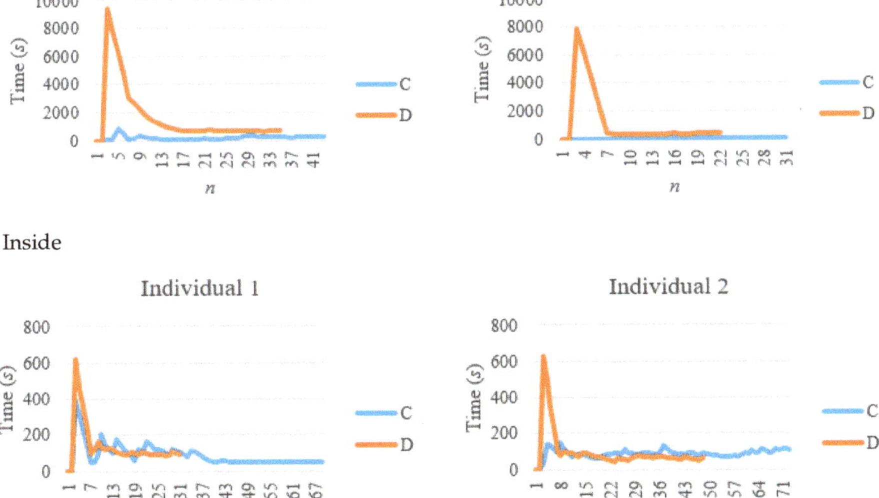

Inside

References

1. Myers, P.; Young, J. Consistent individual behavior: Evidence of personality in black bears. *J. Ethol.* **2018**, *36*, 117–124. [CrossRef]
2. Santicchia, F.; Wauters, L.A.; Dantzer, B.; Westrick, S.E.; Ferrari, N.; Romeo, C.; Palme, R.; Preatoni, D.G.; Martinoli, A. Relationships between personality traits and the physiological stress response in a wild mammal. *Curr. Zool.* **2019**, *66*, 197–204. [CrossRef]
3. Thys, B.; Lambreghts, Y.; Pinxten, R.; Eens, M. Nest defence behavioral reaction norms: Testing life-history and parental investment theory predictions. *R. Soc. Open Sci.* **2019**, *6*, 182180. [CrossRef] [PubMed]
4. Montiglio, P.-O.; Garant, D.; Pelletier, F.; Re´ale, D. Personality differences are related to long-term stress reactivity in a population of wild eastern chipmunks, *Tamias striatus*. *Anim. Behav.* **2012**, *84*, 1071–1079. [CrossRef]
5. Luo, L.; Reimert, I.; de Haas, E.N.; Kemp, B.; Bolhuis, J.E. Effects of early and later life environmental enrichment and personality on attention bias in pigs (*Sus scrofa domesticus*). *Anim. Cogn.* **2019**, *22*, 959–972. [CrossRef]

6. Dingemanse, N.J.; Kazem, A.J.N.; Réale, D.; Wright, J. Behavioural reaction norms: Animal personality meets individual plasticity. *Trends Ecol. Evol.* **2009**, *25*, 82–89. [CrossRef]

7. Wilson, V.; Guenther, A.; Øverli, Ø.; Seltmann, M.W.; Altschul, D. Future directions for personality research: Contributing new insights to the understanding of animal behavior. *Animals* **2019**, *9*, 240. [CrossRef]

8. Clubb, R.; Mason, G.J. Animal welfare: Captivity effects on wide-ranging carnivores. *Nature* **2003**, *425*, 473–474. [CrossRef]

9. Mason, G.J.; Latham, N.R. Can't stop, Won't stop: Is stereotypy a reliable animal welfare indicator? *Anim. Welf.* **2004**, *13*, 57–69.

10. Shepherdson, D.; Lewis, K.D.; Carlstead, K.; Bauman, J.; Perrin, N. Individual and environmental factors associated with stereotypic behavior and fecal glucocorticoid metabolite levels in zoo housed polar bears. *Appl. Anim. Behav. Sci.* **2013**, *147*, 268–277. [CrossRef]

11. Carlstead, K.; Seidensticker, J. Seasonal variation in stereotypic pacing in an american black bear *Ursus americanus*. *Behav. Process.* **1991**, *25*, 155–161. [CrossRef]

12. Ross, S.R. Issues of choice and control in the behaviour of a pair of captive polar bears *Ursus maritimus*. *Behav. Process.* **2006**, *73*, 117–120. [CrossRef] [PubMed]

13. Shyne, A. Meta-analytic review of the effects of enrichment on stereotypic behavior in zoo mammals. *Zoo Biol.* **2006**, *25*, 317–337. [CrossRef]

14. Cless, I.T.; Voss-Hoynes, H.A.; Ritzmann, R.E.; Lukas, K.E. Defining pacing quantitatively: A comparison of gait characteristics between pacing and non-repetitive locomotion in zoo-housed polar bears. *Appl. Anim. Behav. Sci.* **2015**, *169*, 78–85. [CrossRef]

15. Carlstead, K.; Seidensticker, J.; Baldwin, R. Environmental enrichment for zoo bears. *Zoo Biol.* **1991**, *10*, 3–16. [CrossRef]

16. Rose, P.E.; Nash, S.M.; Riley, L.M. To pace or not to pace? A Review of what abnormal repetitive behavior tells us about zoo animal management. *J. Vet. Behav.* **2017**, *20*, 11–21. [CrossRef]

17. Altmann, J. Observational study of behavior: Sampling methods. *Behaviour* **1974**, *49*, 227–267. [CrossRef]

18. Bashaw, M.J.; Kelling, A.S.; Bloomsmith, M.A.; Maple, T.L. Environmental effects on the behavior of zoo-housed lions and tigers, with a case study of the effects of a visual barrier on pacing. *J. Appl. Anim. Welf. Sci.* **2007**, *10*, 95–109. [CrossRef]

19. Pertoldi, C.; Bahrndorff, S.; Novicic, Z.K.; Rohde, P.D. The novel concept of "behavioral instability" and its potential applications. *Symmetry* **2016**, *8*, 135. [CrossRef]

20. Bech-Hansen, M.; Kallehauge, R.M.; Bruhn, D.; Castenschiold, J.H.F.; Gehrlein, J.B.; Laubek, B.; Jensen, L.F.; Pertoldi, C. Effect of landscape elements on the symmetry and variance of the spatial distribution of individual birds within foraging flocks of geese. *Symmetry* **2019**, *11*, 1103. [CrossRef]

21. Clark, F.; King, A.J. A critical review of zoo-based olfactory enrichment. In *Chemical Signals in Vertebrates 11*; Hurst, J.L., Beynon, R.J., Roberts, S.C., Wyatt, T.D., Eds.; Springer: New York, NY, USA, 2008.

22. RStudio Team. *RStudio: Integrated Development for R*; RStudio, Inc.: Boston, MA, USA, 2016.

23. Hammer, Ø.; Harper, D.A.T.; Ryan, P.D. Past: Paleontological statistics software package for education and data analysis. *Palaeontol. Electron.* **2001**, *4*, 1–9.

24. Leys, C.; Ley, C.; Klein, O.; Bernard, P.; Licata, L. Detecting outliers: Do not use standard deviation around the mean, use absolute deviation around the median. *J. Exp. Soc. Psychol.* **2013**, *49*, 764–766. [CrossRef]

25. Zar, J.H. *Biostatistical Analysis*, 4th ed.; Pearson: London, UK; Northern Illinois University: DeKalb, IL, USA, 1999.

26. Yosef, R.; Raz, M.; Ben-Baruch, N.; Shmueli, L.; Kosicki, L.J.Z.; Fratczak, M.; Tryjanowski, P. Directional preferences of dogs' changes in the presence of a bar magnet: Educational experiments in Israel. *J. Vet. Behav.* **2019**, in press. [CrossRef]

27. Rousseeuw, P.J.; Croux, C. Alternatives to the median absolute deviation. *J. Am. Stat. Assoc.* **1993**, *88*, 1273–1283. [CrossRef]

Article

Use of Platelet-Rich Fibrin Associated with Xenograft in Critical Bone Defects: Histomorphometric Study in Rabbits

Paulo Wilson Maia [1], Marcelo Lucchesi Teixeira [1], Luís Guilherme Scavone de Macedo [1], Antonio Carlos Aloise [1], Celio Amaral Passos Junior [1], Juan Manuel Aragoneses [2], José Luis Calvo-Guirado [3] and André Antonio Pelegrine [1,*]

[1] Faculdade São Leopoldo Mandic, Instituto de Pesquisas São Leopoldo Mandic, Campinas 13045-755, Brazil; pwmaia@yahoo.com.br (P.W.M.); marceloltx@gmail.com (M.L.T.); drluismacedo@yahoo.com.br (L.G.S.d.M.); aca.orto@uol.com.br (A.C.A.); cellioamaral@gmail.com (C.A.P.J.)

[2] Department of Dental Research in Universidad Federico Henríquez y Carvajal (UFHEC), 10107 Santo Domingo, Dominican Republic; jmaragoneses@gmail.com

[3] Department of Oral and Implant Surgery, Faculty of Health Sciences, Universidad Católica San Antonio de Murcia (UCAM), 30107 Murcia, Spain; jlcalvo@ucam.edu

* Correspondence: andre.pelegrine@slmandic.edu.br; Tel.: +55-19-981737532

Received: 20 September 2019; Accepted: 11 October 2019; Published: 15 October 2019

Abstract: Platelet-rich fibrin (PRF) is an autologous material used to improve bone regeneration when associated with bone grafts. It affects tissue angiogenesis, increasing the healing process and, theoretically, presenting potential to increase bone neoformation. The aim of this study was to verify, histomorphometrically, the effects of the association of PRF to a xenograft. Twelve adult white New Zealand rabbits were randomly assigned into two groups containing six animals each. After general anesthesia of the animals, two critical defects of 12 mm were created in the rabbit calvaria, one on each side of the sagittal line. Each defect was filled with the following biomaterials: in the control group (CG), xenograft hydrated with saline solution filling one defect and xenograft hydrated with saline solution covered with collagen membrane on the other side; in the test group (TG), xenograft associated with PRF filling the defect of one side and xenograft associated with PRF covered with collagen membrane on the other side. After eight weeks the animals were euthanized and a histomorphometric analysis was performed. The results showed that in the sites that were covered with collagen membrane, there was no statistically significant difference for all the analyzed parameters. However, when comparing the groups without membrane coverage, a statistically significant difference could be observed for the vital mineralized tissue (VMT) and nonmineralized tissue (NMT) parameters, with more VMT in the test group and more NMT in the control group. Regarding the intragroup comparison, the use of the membrane coverage presented significant outcomes in both groups. Therefore, in this experimental model, PRF did not affect the levels of bone formation when a membrane coverage technique was used. However, higher levels of bone formation were observed in the test group when membrane coverage was not used.

Keywords: fibrin; platelets; bone graft; bone regeneration; bone tissue

---

1. Introduction

Among the determining conditions for implant placement in their ideal prosthetic position, the presence of sufficient bone volume is still a determining factor for surgeons [1,2]. In such cases where this volume is insufficient, procedures for bone tissue reconstruction have been described in the literature with satisfactory results by guided bone regeneration technique or onlay block bone grafts. For adequate bone regeneration, the presence of viable bone cells is mandatory and, therefore, if the recipient bed cannot provide these cells to the graft material, it must contain viable bone cells by itself [3]. In this scope, autogenous bone grafts are still considered the gold standard for such procedures, in view of their biological potential (i.e., osteogenic, osteoinductive, and osteoconductive properties). However, the disadvantages associated with the procedure for harvesting grafting material have led professionals to seek new alternatives for bone reconstruction [4]. Time of procedure and postoperative discomfort, such as pain, edema, and bleeding, are some of the reported morbidity issues. Moreover, the amount of tissue available in the donor region and the quality can significantly influence the final outcome of the reconstruction [4]. Due to these factors, the search for substitute materials that could conduct bone formation has increased significantly.

Different biomaterials have been described as alternatives for autogenous bone, such as allogeneic, xenogeneic, and synthetic bone grafts [5]. However, the possibility of immunological reactions and the quality of newly formed tissue have been the subject of discussion due to their low density, compromising, in some clinical situations, the initial stability of implants and consequently, osseointegration [6]. The association of xenogeneic biomaterials with bone marrow-derived stem cells in reconstruction of critical defects has shown good results in experimental animal models with satisfactory vital mineralized tissue formation [7,8]. Similar associations performed in maxillary sinus floor elevation and horizontal bone augmentation procedures demonstrated better results in the formation of a vital mineralized tissue [9,10]. Nowadays, the nanotechnology is also collaborating with the tissue engineering field by developing more adequate scaffolds that can be associated with stem cells and signaling molecules. This triad has a high potential for several types of tissue construction [11]. However, the method of collecting and processing such cells is considered critical and high cost, which does not always make it possible to be performed in a clinical environment.

In contrast with stem cells approaches, platelet-rich fibrin (PRF) represents a simple "chair side" technique of platelet concentrates, first described in the literature to improve tissue regeneration and accelerate the healing process in surgical procedures [12–14]. It consists of a dense fibrin network, after collecting peripheral blood by simple venipuncture, containing leukocytes and platelets, which secrete important growth factors for the regenerative process, such as platelet-derived growth factor (PGDF), vascular endothelial growth factor (VEGF), and transforming growth factor beta (TGF-β). These growth factors act in the processes of angiogenesis, reepithelization, and extracellular matrix formation. Some studies report that such growth factors may be released over a period of up to 28 days. In addition, important adhesive proteins such as fibronectin and vitronectin are also secreted in PRF concentrates [12]. Regardless, studies have shown the possibility of associating PRF with bone substitute biomaterials [13,14]; the association of xenogeneic grafts with PRF in bone reconstructions of critical defects is scarce in the literature. Thus, the aim of this study was to analyze the association of platelet-rich fibrin with xenogeneic graft in critical defects in rabbit calvaria through histomorphometric evaluation. The null hypothesis of the present study is that the use of PRF would not result in higher levels of bone formation.

2. Materials and Methods

Twelve adult male New Zealand rabbits between 10 and 12 months of age, with an average weight of 3.5 kg, were selected. The animals were adapted for the environment and then kept in individual cages with controlled temperature between 18 °C and 20 °C and food and water ad libitum. In all animals, general anesthesia was induced by ketamine (40 mg / kg), midazolam (2 mg/kg), and fentanyl citrate (0.8 mg/kg). Maintenance was performed by a mixture of (isoflurane/N$_2$O [1:1.5%]):oxygen [2/3:1/3] using a pediatric laryngeal mask airway. This study was analyzed and approved by the Research Ethics Committee of the São Leopoldo Mandic Dental School, Campinas, SP, Brazil (process 0191/14).

2.1. Experimental Design

Two circular critical bone defects were performed using a 12 mm external diameter trephine drill (Neodent, Brazil) in the calvaria of the 12 rabbits, totaling 24 defects, distributed in two groups: test group (TG) n = 6, in which the mixture of platelet-rich fibrin (PRF) and xenograft was used as filler; and control group (CG) n = 6, in which only saline solution was used in association with xenograft to fill the defects. The animals were randomly inserted in CG or TG using the tool provided in the website www.randomization.com. The biomaterials used in this study were: Bio-Oss® (Geistlich Biomaterials, Wolhusen, Switzerland) xenograft, a particulate bovine mineral matrix with granules between 0.25 mm and 1 mm; Bio-Gide® (Geistlich Biomaterials, Wolhusen, Switzerland), a porcine resorbable bilaminar collagen membrane; and platelet-rich fibrin, obtained from the centrifugation of the autologous blood of each animal (Figure 1).

Figure 1. Platelet-rich fibrin (PRF) immediately after removal from the tube.

2.2. Preparation and Application of PRF

After the general anesthesia of the rabbits (test group), a 10 ml blood sample was collected from the ear vein of each animal with a 40 × 10 needle and a 20 ml luer syringe (BD® Sigma-Aldrich Brasil Ltda, 0800-7277292, Sao Paulo, Brazil). The sample tubes without anticoagulant were arranged in the opposite position in the Intra-Spin L-PRF Centrifuge (Intralock, Germany), which was used for 10 min at a speed of 3000 rpm. The fibrin clot was removed from the tube with the aid of clinical tweezers (Figure 2), and after its removal the red portion was separated and discarded and the white portion was left for 10 min on a perforated metal surface to drain the remaining liquids. Due to the fact that it was a solid material, the obtained PRF was chopped into approximately 3 mm portions using surgical scissors so that it could be mixed with the particulate xenograft (Figure 3).

Figure 2. Xenograft mixed with PRF.

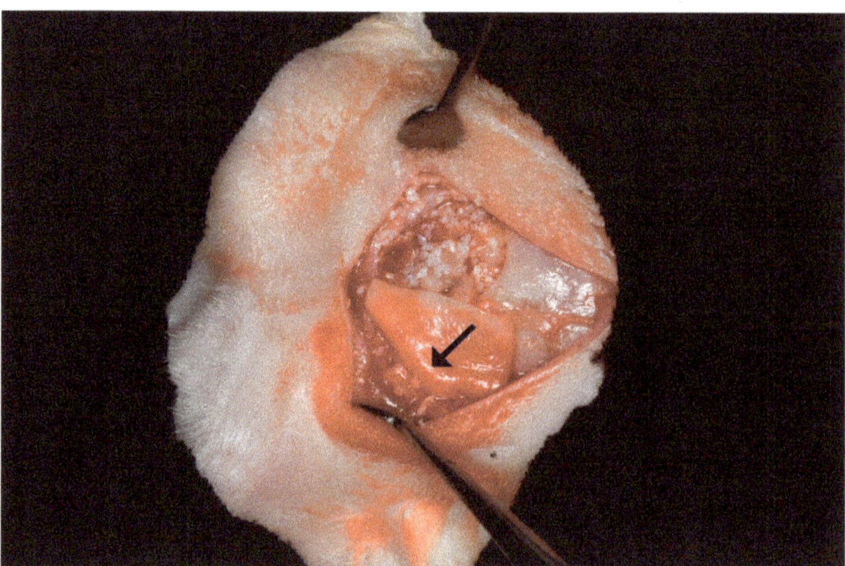

Figure 3. Bone defects filled with the biomaterial covered (rabbit's right side; arrow) or not covered (rabbit's left side) with a barrier membrane.

2.3. Biopsies Collection

All animals were euthanized by anesthetic overdose eight weeks after surgery. To collect the parietal bone samples, incision and a mucoperiosteal flap were performed, followed by the removal of approximately 20 mm^2 of bone containing the grafted healed area. The osteotomy was performed with a 701 drill (Tri-Hawk, Morrisburg, ON, Canada) adapted to the electric motor handpiece used in the previous phase.

2.4. Histologic and Histomorphometric Analysis

The samples were fixed in 10% buffered formalin for 48 h. They were then treated with 10% EDTA solution for decalcification for one hour. Histological analysis was performed on the central portion of the bone biopsies and after histological treatment, 7 μm sections of each specimen were obtained and stained with hematoxylin–eosin and examined by light microscopy (Figure 4A,B and Figure 5A,B). Microscopic images were captured using a CCD digital camera (RT Color; Diagnostic Instruments, Sterling Heights, MI, USA). Images were analyzed using Image Pro Plus 4.5 software for Windows (Media Cybernetics, San Diego, CA, USA). The evaluated parameters were: (1) nonvital mineralized tissue (NVMT); (2) vital mineralized tissue (VMT); and (3) nonmineralized tissue (NMT). Connective tissue, bone marrow, blood vessels, and adipose tissue (i.e., all tissues that cannot be considered as a mineralized tissue) were considered NMT. Mineralized tissues (stained in darker pink) were considered vital (VMT) when osteocytes were present and nonvital (NVMT) when osteocytes were absent. All results were measured in μm^2 and expressed as a percentage of the total area.

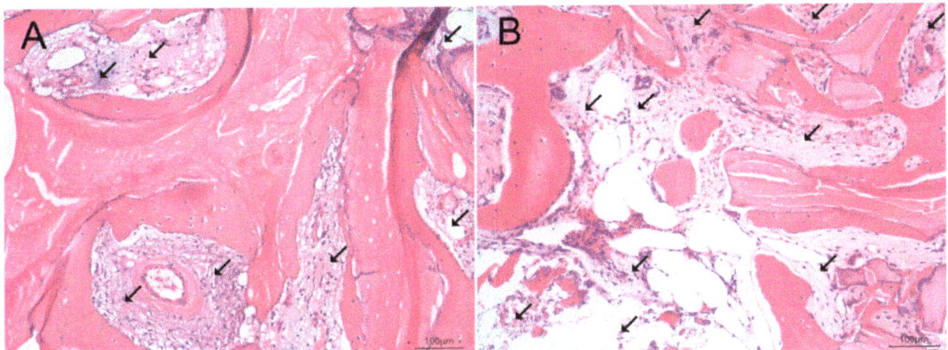

Figure 4. (**A**) Histological view of a test group subject with membrane coverage specimen (200× magnification). Note the new bone formation in between biomaterial particles and the lower amount of nonmineralized tissue (arrows = nonmineralized tissue). (**B**) Histological view of a control group subject without membrane coverage specimen (200× magnification). Note the higher amount of nonmineralized tissue (arrows = non mineralized tissue).

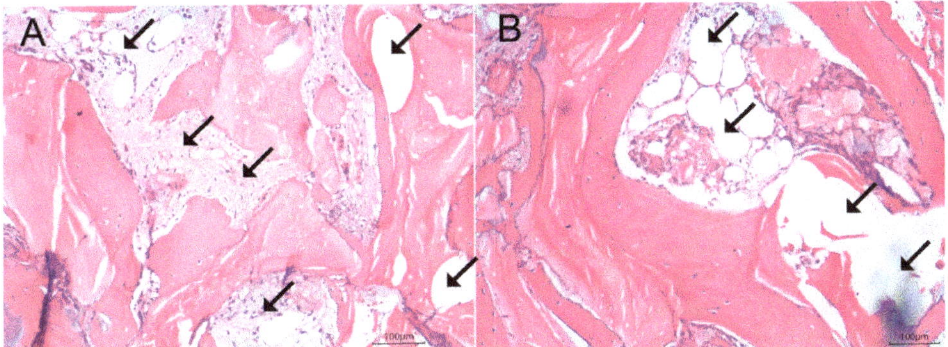

Figure 5. (**A**) Histological view of a test group subject without membrane coverage specimen (200× magnification) (arrows = nonmineralized tissue). (**B**) Histological view of a control group subject with membrane coverage specimen (200× magnification) (arrows = nonmineralized tissue). Note the similar amount of nonmineralized tissue between the two samples.

2.5. Statistical Analysis

The software SPSS-V17 (SPSS Inc. 233, Chicago, IL, USA) was used to analyze all quantitative data. The Kruskall–Wallis test was used to compare groups for each TMNV, TMV, and TNM variable. The Wilcoxon test was used to compare the results obtained with and without membrane coverage. The value of $p \leq 0.05$ indicated statistical significance.

3. Results

Histomorphometry

Comparing the test and control groups (intergroup comparison) in sites that were covered with collagen membrane, there was no statistically significant difference for all the analyzed parameters. However, when comparing the groups without membrane coverage, a statistically significant difference could be observed for the vital mineralized tissue (VMT) and nonmineralized tissue (NMT) parameters, with more VMT in the test group and more NMT in the control group. Regarding the intragroup comparison, the use of the membrane coverage presented significant outcomes in both groups, with a higher level of VMT and lower level of NMT (Table 1).

Table 1. The results of the histomorphometric analysis.

Tissue	With Membrane		*p*-Value	Without Membrane		*p*-Value
	Control Group	Test Group		Control Group	Test Group	
NVMT	13.50 ± 0.09	17.88 ± 7.57	0.2971	13.09 ± 0.06	16.59 ± 3.54	0.0542
VMT	13.29 ± 0.16	13.88 ± 5.68	0.5211	6.57 ± 0.08	9.67 ± 0.60	0.0039 *
NMT	73.21 ± 0.12	68.24 ± 5.05	0.0776	80.34 ± 0.05	73.74 ± 3.48	0.0038 *

Table 1—Inter-group analysis: statistical comparison of mean values (in %) between the TG (n = 6) and CG (n = 6) groups. VMT: vital mineralized tissue; NVMT: nonvital mineralized tissue; NMT: nonmineralized tissue; TG (test group); CG (control group). Statistically significant *, $p \leq 0.05$ (Kruskal–Wallis test).

4. Discussion

Autogenous bone grafts are still considered the biological "gold standard" for the reconstruction of critical defects, but the morbidity associated with material collection is the major disadvantage factor. Thus, the search for increased osteogenic capacity of xenogeneic bone substitute biomaterials has been verified in several preclinical and clinical studies [7,8,15]. These studies were conducted to contribute with scientific evidence that xenogeneic biomaterials, although acellular, could acquire osteogenicity when associated with biological tissues and/or cells, and not just osteoconduction [16]. Therefore, some highly cellularized biological tissues, such as fresh and centrifuged bone marrow, as well as bone marrow and adipose tissue stem cells, have been associated with xenogeneic bone grafts to analyze the amount of vital mineralized tissue (VMT), that is, the newly formed bone, in experimental animal models—more specifically, models of critical defects in rabbit calvaria, similar methodology to that used in this study [7,8,17]. Such studies have shown results of increased amount of newly formed tissue due to the association of fresh biological tissue or the use of adult mesenchymal stem cells with xenogeneic biomaterial. However, despite promising published results, a relevant problem still lies in the degree of morbidity associated with bone marrow collection, in addition to its preparation, which must be performed within a hospital environment, increasing the cost and working time [8]. In contrast, the use of venous blood products has some advantages, such as easy collection, low cost, and possibility of being performed in a clinic environment [18]. Platelet-rich fibrin (PRF) is a peripheral blood derivative obtained by centrifugation and consists of a dense matrix of fibrin, rich in growth factors, leukocytes, and adhesive proteins, important in the healing process, accelerating vascularization and subsequent tissue regeneration. Despite these relevant properties, the isolated use of PRF has not shown significant results in previous studies, which is major reason the present study used it in combination with a mineralized xenogeneic biomaterial in critical defects in rabbit calvaria, a methodology that had not yet been verified with caution in the literature with the use of PRF.

Based on methodology previously published in an experimental animal model by Pelegrine et al. (2014) [7] and Aloise et al. (2015) [8], in this study we decided to use a porcine collagen membrane to cover one of the two critical defects produced in animal calvaria, with the intention of establishing an analysis from the perspective of guided bone regeneration [19]. Thus, it was possible to establish correlations between the usage, or not, of PRF, as well as the use, or not, of collagen membrane (i.e., guided bone regeneration concept). In this study the values for VMT in the control group (CG) and test group (TG) on the side without the membrane coverage were 6.56% ± 0.08% and 9.52% ± 0.60%, respectively, with statistical difference ($p < 0.05$). Values for the membrane-covered side were 13.29% ± 0.16% and 13.88% ± 5.68% respectively for CG and TG, with no statistical difference ($p > 0.05$). For NMT, the values for the control group (CG) and test group (TG) on the side without the membrane cover were 80.34% ± 0.05% and 73.74% ± 3.48%, respectively, with statistical difference ($p < 0.05$). Values for the membrane-coated side were 73.21% ± 0.12% and 68.24% ± 5.05%, respectively, with no statistical difference between them ($p > 0.05$). The level of NMT was higher in CG without membrane coverage and lower in TG with membrane coverage, which can be clearly noticed in Figures 4 and 5. These results demonstrate that the addition of PRF cell components to particulate bone grafts allows an increase in the amount of newly formed bone and a decrease in nonmineralized tissues, only when a membrane related to the guided bone regeneration technique is not used. However, when the membrane was used, the usage of PRF did not reflect any change in the mineralization pattern, corroborating the findings of Bolukbasi et al. (2015) [13]. A possible hypothesis for this finding is that PRF in the control group may have performed some bioactive barrier function to peripheral fibroblasts, but without overcoming the barrier role of the collagen membrane used in this study. Therefore, it is important to state that the membrane technique may be the choice used in therapeutics.

When the values of VMT obtained in this study are compared with the results obtained by Aloise et al. (2015) [8] and Pelegrine et al. (2014) [7], who made use of the same experimental model of the present study (i.e., critical bone defect produced in rabbit calvaria), the values of this study are lower. This may indicate that the stimulation caused by PRF is less than that generated by fresh bone marrow, centrifuged bone marrow, and bone marrow stem cells, which may be linked to the fact that bone marrow and bone marrow-derived stem cells have higher affinity for bone tissue than venous blood cells used for PRF production. In this study, the analysis of the amount of nonvital mineralized tissue (NVMT), which represents the residual particles of the xenogeneic graft, showed no statistical difference between CG and TG, with or without collagen membrane coverage. This shows that the level of resorption of xenogeneic graft has not significantly changed, regardless of whether or not the concept of guided bone regeneration was used. This finding may be related to the fact that the biomaterial used in this study was a bovine hydroxyapatite, which presents a very slow resorption pattern [20]. However, it is reasonable to consider that in the test group, probably, the amount of bovine hydroxyapatite used was smaller, since PRF seems to occupy significant volume between bone particles (data not shown).

5. Conclusions

Based in this experimental model we can conclude that:

(1) Platelet-rich fibrin does not result in higher levels of bone formation when a guided bone regeneration technique (i.e., with membrane coverage) is used.
(2) The usage of the collagen membrane has a synergistic effect on bone healing when associated with a xenograft.
(3) Platelet-rich fibrin may increase the level of newly formed bone only in bone grafting procedures using xenograft without collagen membrane coverage.

Author Contributions: Conceptualization, P.W.M. and A.C.A.; data curation M.L.T. and A.A.P.; formal analysis P.W.M. and A.C.A.; funding acquisition, M.L.T. and A.A.P.; investigation P.W.M. and A.A.P.; methodology, P.W.M. and A.A.P.; project administration A.C.A. and L.G.S.d.M.; resources M.L.T. and C.A.P.J.; software, P.W.M., J.M.A., and L.G.S.d.M.; supervision M.L.T. and J.M.A.; validation, J.L.C.-G. and J.M.A.; visualization A.C.A. and A.A.P.; writing—original draft preparation, A.A.P. and J.M.A.; writing—review and editing J.L.C.-G. and A.A.P.

Funding: This research received no external funding.

Conflicts of Interest: The authors declare no conflict of interest.

References

1. Chrcanovic, B.; Albrektsson, T.; Wennerberg, A. Bone Quality and Quantity and Dental Implant Failure: A Systematic Review and Meta-analysis. *Int. J. Prosthodont.* **2017**, *30*, 219–237. [CrossRef] [PubMed]

2. Bagher, Z.; Rajaei, F.; Shokrgozar, M. Comparative Study of Bone Repair Using Porous Hydroxyapatite/β-Tricalcium Phosphate and Xenograft Scaffold in Rabbits with Tibia Defect. *Iran. Biomed. J.* **2012**, *16*, 18–24. [PubMed]

3. Stevenson, S.; Emery, S.E.; Goldberg, V.M. Factors Affecting Bone Graft Incorporation. *Clin. Orthop. Relat. Res.* **1996**, *324*, 66–74. [CrossRef] [PubMed]

4. Gultekin, B.A.; Bedeloglu, E.; Kose, T.E.; Mijiritsky, E. Comparison of Bone Resorption Rates after Intraoral Block Bone and Guided Bone Regeneration Augmentation for the Reconstruction of Horizontally Deficient Maxillary Alveolar Ridges. *BioMed Res. Int.* **2016**, *2016*, 4987437. [CrossRef] [PubMed]

5. Titsinides, S.; Agrogiannis, G.; Karatzas, T. Bone grafting materials in dentoalveolar reconstruction: A comprehensive review. *Jpn. Dent. Sci. Rev.* **2019**, *55*, 26–32. [CrossRef] [PubMed]

6. De Lacerda, P.E.; Pelegrine, A.A.; Teixeira, M.L.; Montalli, V.A.M.; Rodrigues, H.; Napimoga, M.H. Homologous transplantation with fresh frozen bone for dental implant placement can induce HLA sensitization: A preliminary study. *Cell Tissue Bank.* **2016**, *17*, 465–472. [CrossRef] [PubMed]

7. Pelegrine, A.A.; Aloise, A.C.; Zimmermann, A.; de Mello E Oliveira, R.; Ferreira, L.M. Repair of critical-size bone defects using bone marrow stromal cells: A histomorphometric study in rabbit calvaria. Part I: Use of fresh bone marrow or bone marrow mononuclear fraction. *Clin. Oral Implant. Res.* **2014**, *25*, 567–572. [CrossRef] [PubMed]

8. Aloise, A.C.; Pelegrine, A.A.; Zimmermann, A.; Oliveira, R.D.M.E.; Ferreira, L.M. Repair of critical-size bone defects using bone marrow stem cells or autogenous bone with or without collagen membrane: A histomorphometric study in rabbit calvaria. *Int. J. Oral Maxillofac. Implant.* **2015**, *30*, 208–215. [CrossRef] [PubMed]

9. Pasquali, P.J.; Teixeira, M.L.; De Oliveira, T.A.; De Macedo, L.G.S.; Aloise, A.C.; Pelegrine, A.A. Maxillary Sinus Augmentation Combining Bio-Oss with the Bone Marrow Aspirate Concentrate: A Histomorphometric Study in Humans. *Int. J. Biomater.* **2015**, *2015*, 121286. [CrossRef] [PubMed]

10. De Oliveira, T.A.; Aloise, A.C.; Orosz, J.E.; Oliveira, R.D.M.E.; De Carvalho, P.; Pelegrine, A.A. Double Centrifugation Versus Single Centrifugation of Bone Marrow Aspirate Concentrate in Sinus Floor Elevation: A Pilot Study. *Int. J. Oral Maxillofac. Implant.* **2016**, *31*, 216–222. [CrossRef] [PubMed]

11. Chieruzzi, M.; Pagano, S.; Moretti, S.; Pinna, R.; Milia, E.; Torre, L.; Eramo, S. Nanomaterials for Tissue Engineering in Dentistry. *Nanomaterials* **2016**, *6*, 134. [CrossRef] [PubMed]

12. Dohan, D.M.; Choukroun, J.; Diss, A.; Dohan, S.L.; Dohan, A.J.; Mouhyi, J.; Gogly, B. Platelet-rich fibrin (PRF): A second-generation platelet concentrate. Part I: Technological concepts and evolution. *Oral Surg. Oral Med. Oral Pathol. Oral Radiol. Endodontol.* **2006**, *101*, e37–e44. [CrossRef] [PubMed]

13. Bolukbasi, N.; Ersanli, S.; Keklikoglu, N.; Basegmez, C.; Özdemir, T. Sinus augmentation with platelet-rich fibrin in combination with bovine bone graft versus bovine bone graft in combination with collagen membrane. *J. Oral Implant.* **2015**, *41*, 586–595. [CrossRef] [PubMed]

14. Borie, E.; Oliví, D.G.; Orsi, I.A.; Garlet, K.; Weber, B.; Beltrán, V.; Fuentes, R. Platelet-rich fibrin application in dentistry: A literature review. *Int. J. Clin. Exp. Med.* **2015**, *8*, 7922–7929. [PubMed]

15. Pelegrine, A.A.; Teixeira, M.L.; Sperandio, M.; Almada, T.S.; Kahnberg, K.E.; Pasquali, P.J.; Aloise, A.C. Can bone marrow aspirate concentrate change the mineralization pattern of the anterior maxilla treated with xenografts? A preliminary study. *Contemp. Clin. Dent.* **2016**, *7*, 21–26. [CrossRef] [PubMed]

16. Lucarelli, E.; Beretta, R.; Dozza, B.; Tazzari, P.; O'Connell, S.; Ricci, F.; Pierini, M.; Squarzoni, S.; Pagliaro, P.; Oprita, E.; et al. A recently developed bifacial platelet-rich fibrin matrix. *Eur. Cells Mater.* **2010**, *20*, 13–23. [CrossRef]

17. Zimmermann, A.; Pelegrine, A.A.; Peruzzo, D.; Martinez, E.F.; Oliveira, R.D.M.E.; Aloise, A.C.; Ferreira, L.M. Adipose mesenchymal stem cells associated with xenograft in a guided bone regeneration model: A histomorphometric study in rabbit calvaria. *Int. J. Oral Maxillofac. Implant.* **2015**, *30*, 1415–1422. [CrossRef] [PubMed]

18. Nacopoulos, C.; Dontas, I.; Lelovas, P.; Galanos, A.; Vesalas, A.-M.; Raptou, P.; Mastoris, M.; Chronopoulos, E.; Papaioannou, N. Enhancement of Bone Regeneration with the Combination of Platelet-Rich Fibrin and Synthetic Graft. *J. Craniofacial Surg.* **2014**, *25*, 2164–2168. [CrossRef] [PubMed]

19. Yoon, J.-S.; Lee, S.-H.; Yoon, H.-J. The influence of platelet-rich fibrin on angiogenesis in guided bone regeneration using xenogenic bone substitutes: A study of rabbit cranial defects. *J. Cranio-Maxillofac. Surg.* **2014**, *42*, 1071–1077. [CrossRef] [PubMed]

20. Pistilli, R.; Felice, P.; Piatelli, M.; Nisii, A.; Barausse, C.; Esposito, M. Blocks of autogenous bone versus xenografts for the reha- bilitation of atrophic jaws with dental implants: Preliminary data from a pilot randomised controlled trial. *Eur. J. Oral Implantol.* **2014**, *7*, 153–171. [PubMed]

Article

Effect of Landscape Elements on the Symmetry and Variance of the Spatial Distribution of Individual Birds within Foraging Flocks of Geese

Mads Bech-Hansen [1,*,†], **Rune M. Kallehauge** [1,†], **Dan Bruhn** [1], **Johan H. Funder Castenschiold** [1], **Jonas Beltoft Gehrlein** [1], **Bjarke Laubek** [2], **Lasse F. Jensen** [2] **and Cino Pertoldi** [1,3]

[1]   Department of Chemistry and Bioscience, Aalborg University, Fredrik Bajers Vej 7H, DK-9220 Aalborg Øst, Denmark
[2]   Vattenfall Vindkraft A/S, Jupitervej 6–2nd floor, DK-6000 Kolding, Denmark
[3]   Aalborg Zoo, Mølleparkvej 63, DK-9000 Aalborg, Denmark
*   Correspondence: madsbechhansen@gmail.com; Tel.: +45-40-266-403
†   Both authors contributed equally.

Received: 23 July 2019; Accepted: 23 August 2019; Published: 2 September 2019

Abstract: Behavioural instability is a newly coined term used for measuring asymmetry of bilateral behavioural traits as indicators of genetic or environmental stress. However, this concept might also be useful for other types of data than bilateral traits. In this study, behavioural instability indices of expected behaviour were evaluated as an indicator for environmental stress through the application of aerial photos of foraging flocks of geese. It was presumed that geese would increase anti-predator behaviour through the dilution effect when foraging near the following landscape elements: wind turbines, hedgerows, and roads. On this presumption, it was hypothesized that behavioural instability of spatial distribution in flocks of geese could be used as indicators of environmental stress. Asymmetry in spatial distribution was measured for difference in flock density across various distances to disturbing landscape elements through the following indices; behavioural instability of symmetry and behavioural instability of variance. The behavioural instability indices showed clear tendencies for changes in flock density and variance of flock density for geese foraging near wind turbines, hedgerows, and roads indicating increasing environmental stress levels. Thus, behavioural instability has proven to be a useful tool for monitoring environmental stress that does not need bilateral traits to estimate instability but can be applied for indices of expected behaviour.

Keywords: environmental stress; behavioural instability; biomonitoring tool; unmanned aerial vehicles (UAVs); disturbance

1. Introduction

1.1. Behavioural Instability

A study by Pertoldi et al. [1] coined the term behavioural instability as an indicator of genetic or environmental stress based on asymmetry in a bilateral behavioural trait, e.g., clockwise and counter-clockwise directional movement. The authors applied the conventional indices used for the estimation of developmental instability in directional movements. However, the concept of behavioural instability can be considered in a broader sense [1], and the concept might prove to be a viable biomonitoring tool in other fields of research, e.g., studies on environmental impact on wildlife. Furthermore, behavioural instability might be useful for other types of data with an expectation of symmetry other than bilateral traits.

1.2. Environmental Stressors of Geese

An important factor influencing the habitat use of water birds is the potential disturbance by landscape elements such as wind turbines [2], hedgerows, and roads [3,4]. Disturbance might be caused by anthropogenic activity or these landscape elements offering hiding spots for predators [5–7], which essentially can result in displacement and habitat loss. Presuming geese foraging near disturbing landscape elements experience increased stress levels as a result of the potential increased predation risk, anti-predator behaviour through the dilution effect would increase [8]. The dilution effect is defined as increasing group size or density as a strategy for reducing individual risk of being targeted by a predator at the cost of reduced foraging efficiency, e.g., due to increased competition [8].

1.3. Behavioural Instability of Symmetry

Based on the presumption of geese increasingly relying on the dilution effect when foraging near disturbing landscape elements, it would be possible to analyse stress as a function of flock densities at various distances to these landscape elements. Thus, it would be possible to show an asymmetrical distribution by plotting the density of the birds versus the distance to an obstacle in a linear regression as illustrated in Figure 1b. If the obstacle is inducing anti-predator behaviour, the slope (a) of the linear regression will become negative (a < 0) indicating a decreasing flock density with distance. Deviations from a symmetrical distribution (a ≠ 0) will be considered as an estimator of behavioural instability and will be referred to as behavioural instability of symmetry (BSYM).

1.4. Behavioural Instability of Variance

Pertoldi et al. [1] also noted that different genotypes have different perceptions of stressors, and that in a suboptimal environment the perception of stress will become more differentiated among genotypes [1]. If this finding is valid for the way in which the birds behave in the presence of stress, then heterogeneity can be expected where the average distance of the residuals from the regression line is not constant at different distances from the landscape element as illustrated in Figure 1c. This variation can be considered as another estimator of behavioural instability, with a higher variance of the residuals being interpreted as a smaller capacity to predict the behaviour of the individuals in the presence of stressors. This estimator is later referred to as deviation from a behavioural instability of variance (BVAR).

1.5. Aims of the Investigation

Based on the mentioned theories, we hypothesized that behavioural instability indices of expected behaviour can be used as a tool for measuring environmental stresses in wildlife. This will be evaluated through the analysis of the behavioural instability of the spatial distribution of flocks of geese relying on two different measures: behavioural instability of symmetry (BSYM) (Figure 1b) and behavioural instability of variance (BVAR) (Figure 1c). These indices were tested using aerial photos of flocks of geese foraging (Figure 1b) near the following landscape elements; wind turbines, hedgerows, and roads, which are all known to cause the disturbance of geese.

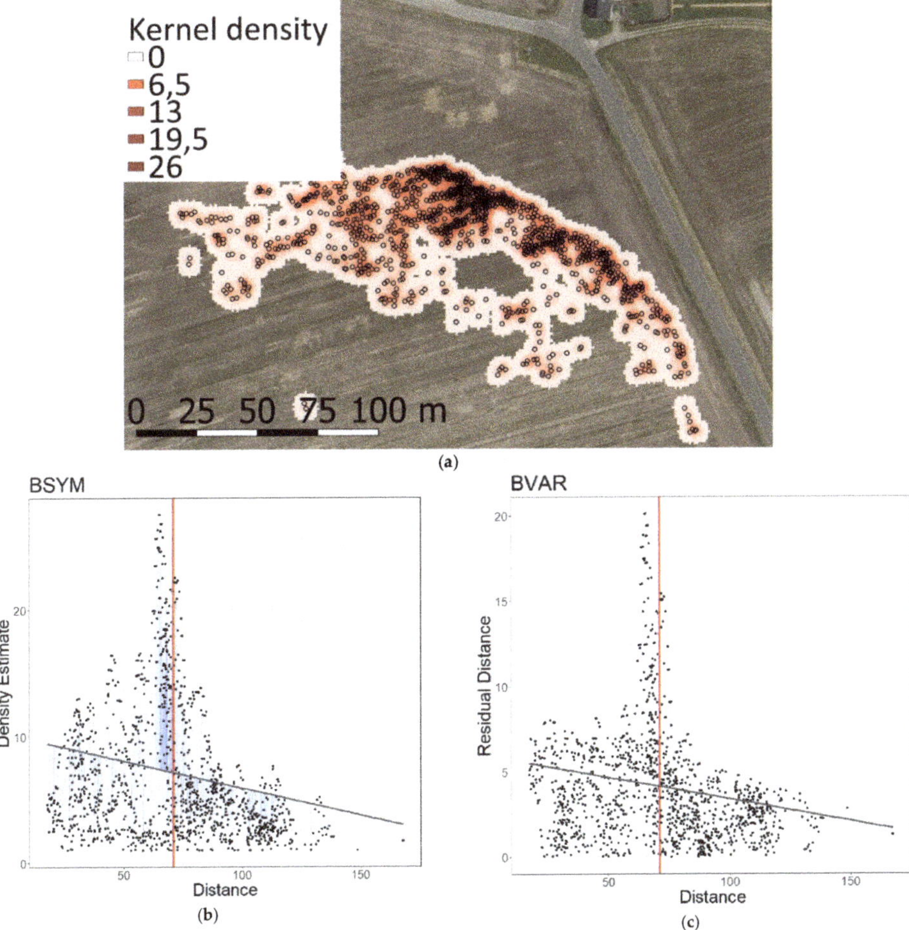

Figure 1. Overview of behavioural instability indices used to monitor one flock of pink-footed geese foraging near a road: (**a**) Aerial photos of a flock of individual geese are georeferenced in geographic information systems (GIS) and geographic positions of individual geese are geotagged (illustrated as circles). Flock density is quantified by Kernel density estimates using the geographic positions of identified birds. Kernel density estimates are illustrated as white (low density) to dark red (high density); (**b**) Density estimates of individuals as a function of distance to the nearest landscape element, in this case, roads. The mean distance is illustrated as a red vertical line. A linear regression of density (LRD): Density Estimate = a × Distance + b (illustrated as a grey line) is fitted to measure behavioural instability through asymmetry of density between distances left and right of the mean distance. Symmetrical density would result in a non-significant slope (a = 0). Contrarily, an anti-predator behaviour induced by the road would yield a negative slope (a < 0). Residual distances between observed densities and LRD is illustrated as blue lines; (**c**) Residual distances of density (Figure 1b) as a function of distance to the nearest landscape elements, in this case roads. A linear regression of residuals (LRR): Residual = a × Distance + b (illustrated as a grey line) is fitted to measure behavioural instability through asymmetry of variance in density between distances left and right of the mean distance. A constant variance in density across distance would result in a non-significant slope (a = 0). Contrarily, increased variance would yield a negative slope (a < 0).

2. Materials and Methods

Aerial photos of foraging geese were used to evaluate changes in anti-predator behaviour by examining the spatial distribution of flocks of different distances to wind turbines, hedgerows, and roads. This was done to evaluate the two indices: behavioural instability of symmetry (BSYM), and behavioural instability of variance (BVAR), as indicators of environmental stressors. The studied species were barnacle goose (*Branta leucopsis*) (19 flocks, with a total of 18,925 individuals), pink-footed goose (*Anser brachyrhynchus*) (23 flocks, with a total of 26,313 individuals), and greylag goose (*Anser anser*) (5 flocks, with a total of 353 individuals).

Spatial distribution was measured based on distances of individuals to said landscape elements, as well as from flock density estimates extracted from geographic information systems (GIS) (Figure 1a). Thus, differences in spatial distribution will be tested between flocks of geese are observed at different distances from the different landscape elements, through the analysis of behavioural instability of symmetry (BSYM) (Figure 1b) and behavioural instability of variance (BVAR) (Figure 1c).

2.1. Data Collection

In the period March 2017 to February 2018 aerial photos of foraging flocks of geese were collected in Northern Jutland, Denmark, mainly around Klim and Nørrekær Enge wind farm, using a UAV of the model DJI Phantom 4 Pro Quadcopter. Data were collected during daytime on 26 different days, chosen based on the season (winter migration of geese) and weather forecast (no rain and low wind speed). All UAV overflights were performed at an altitude of 100 m either flown manually using the DJI GO 4 Drone application (Version 4.0.6), with the video camera capturing aerial photos vertical downwards in 4K resolution or flown automatically using DroneDeploy (Version 2.69). Additionally, data collection using the UAV were conducted following the recommended flight altitude (100 m) and take-off distance from the studied flocks (~500 m) suggested by Bech-Hansen et al. [9] as a means to prevent initial disturbance of the birds.

2.2. Data Extraction

Aerial photos were imported into QGIS (Version 2.8.20) using the geo-referencing plugin GDAL (Version 3.1.9) as large ortho-mosaics created using Autostitch (Demo version) or DroneDeploy limited free services. Birds were then identified, given UTM coordinates, and, if possible, given a species-specific id. A total of 45,591 birds were identified from 47 flocks of geese. Aerial photos were not corrected for barrel distortion effects as it was considered of minor influence.

For each flock, a density heatmap was produced in QGIS from the location of identified birds using the Heatmap plugin (Version 0.2), based on the Kernel density estimation [10], with radius set to 5 m, and cell size x and y set to 1 m each, thus, resulting in comparable density estimates across all studied flocks. Density estimates for each identified bird were extracted using the QGIS plugin Point Sampling Tool (Version 0.4.1), while the distance between the identified birds and nearest wind turbine, hedgerow, and road, were measured using the NNJoin plugin (Version 1.3.1) in QGIS.

2.3. Data Analysis

All analyses were conducted based on bird distances to the three landscape elements: wind turbines, hedgerows, and roads. Based on previous studies, birds at distances of more than 600 m from the nearest associated landscape structure were assumed not to be affected by these structures [2–4,6] and were therefore excluded from the analysis.

Density estimates of all individual birds were plotted as a function of bird distances to nearest landscape element and fitted with a linear regression (Figure 1b), which is, henceforth, referred to as linear regression of density (LRD) [Equation (1)].

$$\text{Density Estimate} = a \times \text{Distance} + b \tag{1}$$

An LRD with negative significant slope (a < 0) indicates an increase in flock density in the direction of the studied landscape element and vice versa. A negative slope also indicates an asymmetrical distribution of the flock density and consequently an increased BSYM in the direction of the studied landscape element.

The absolute values of the residual distances of the LRD [Equation (2)] as previously mentioned, were plotted versus the distance to the obstacles in a linear regression of residuals (LRR) [Equation (3)] (Figure 1c).

$$\text{Residual} = \text{Observed density} - \text{predicted density (LRD)} \tag{2}$$

$$\text{Residual} = a \times \text{Distance} + b \tag{3}$$

A slope of LRR different from 0 (a ≠ 0) indicates heterogeneity of the residuals. A negative slope (a < 0) indicates an increasing variance of the residuals' absolute distance from LRD in the direction of the studied landscape element and vice versa. Therefore, an increased variance can be considered as an increased BVAR in the direction of the studied landscape element and vice versa.

The determination coefficients (r^2), the slope and the intercept of all LRDs and LRRs have been estimated as well as significance for each slope (later noted as asterisks; *: $p < 0.05$, **: $p < 0.01$, ***: $p < 0.001$).

3. Results

3.1. Correlation between Distance to Landscape Elements and Behavioural Instability of Symmetry (BSYM)

3.1.1. Wind Turbines

The barnacle geese and pink-footed geese showed a significant increase of BSYM with decreasing distance to the wind turbines ($r^2 = 20.53\%$) *** and ($r^2 = 10.51\%$) *** respectively (Table 1 and Figure 2a,d). The greylag geese showed the same trend, but the regression was not significant ($r^2 = 0.45\%$, $p > 0.05$) (Table 1 and Figure 2g).

3.1.2. Roads

The greylag geese showed a significant increase of BSYM with decreasing distance to the roads ($r^2 = 8.08\%$) *** (Table 1 and Figure 2h). Whereas, barnacle geese and pink-footed geese showed (with $r^2 < 5\%$) the opposite trend with a significant decrease of BSYM with decreasing distance to the roads ($r^2 = 0.04\%$) *** and ($r^2 = 1.01\%$) *** respectively (Table 1 and Figure 2b,e).

3.1.3. Hedgerows

Barnacle geese and pink-footed geese showed a significant (with $r^2 < 5\%$) increase of BSYM with decreasing distance to the hedgerows ($r^2 = 1.46\%$) *** and ($r^2 = 2.13\%$) *** respectively (Table 1 and Figure 2c,f).

The greylag geese showed (with a non-significant regression and $r^2 < 5\%$) the opposite trend with a significant decrease of BSYM with decreasing distance to the hedgerows ($r^2 = 0.18\%$, $p > 0.05$) (Table 1 and Figure 2i).

3.2. Correlation between Distance to Landscape Elements and Behavioural Instability of Variance (BVAR)

3.2.1. Wind Turbines

The barnacle geese and pink-footed geese showed a significant increase of BVAR with decreasing distance to the wind turbines ($r^2 = 14.69\%$) *** and ($r^2 = 22.18\%$) *** respectively (Table 1 and Figure 3a,d). Whereas, the greylag geese showed (with $r^2 < 5\%$ and $p > 0.05$) the opposite trend with a decrease of BVAR with decreasing distance to the wind turbines (Table 1 and Figure 3g).

Table 1. Species: Barnacle goose (BG), pink-footed goose (PINK), and greylag goose (GREY); distances from the landscape elements: wind turbines (dwi), roads (dro), and hedgerows (dhe), n = number of measurements. Coefficient of determination (r^2) slope (a), intercept (b) and significance level (p) of both linear regression of density (LRD) (which regress the density estimates of individual birds as a function of bird distances to nearest landscape element) and of linear regression of residuals (LRR) (which regress the absolute values of residual distance from LRD as a function of bird distances to nearest landscape element). The r^2 values above 5% ($r^2 > 5\%$) are in bold.

Species Dist. from Obstacles		n (Number of Measurements)	LRD r^2	LRD a & b		LRD p	LRR r^2	LRR a & b		LRR p
BG	dwi	4872	20.53%	Slope a: Intercept b:	−0.052 34.182	***	14.69%	Slope a: Intercept b:	−0.023 15.038	***
	dro	18925	0.04%	Slope a: Intercept b:	0.001 10.258	**	2.35%	Slope a: Intercept b:	−0.004 5.394	**
	dhe	18925	1.46%	Slope a: Intercept b:	−0.007 11.610	***	5.20%	Slope a: Intercept b:	−0.008 5.968	***
PINK	dwi	4894	10.51%	Slope a: Intercept b:	−0.035 28.681	***	22.18%	Slope a: Intercept b:	−0.026 17.196	***
	dro	26394	1.01%	Slope a: Intercept b:	−0.008 8.744	***	2.64%	Slope a: Intercept b:	−0.008 6.301	***
	dhe	26313	2.13%	Slope a: Intercept b:	0.008 11.588	***	2.74%	Slope a: Intercept b:	−0.005 5.734	***
GREY	dwi	361	0.45%	Slope a: Intercept b:	−0.002 3.720	n.s.	0.65%	Slope a: Intercept b:	0.001 1.090	n.s.
	dro	458	8.08%	Slope a: Intercept b:	−0.006 4.176	***	10.67%	Slope a: Intercept b:	−0.004 1.910	***
	dhe	353	0.18	Slope a: Intercept b:	0.001 2.873	n.s.	0.46%	Slope a: Intercept b:	−0.001 1.22	n.s.

$p < 0{,}05 = *$, $p < 0{,}01 = **$, $p < 0.0001 = ***$, and n.s. = non-significant.

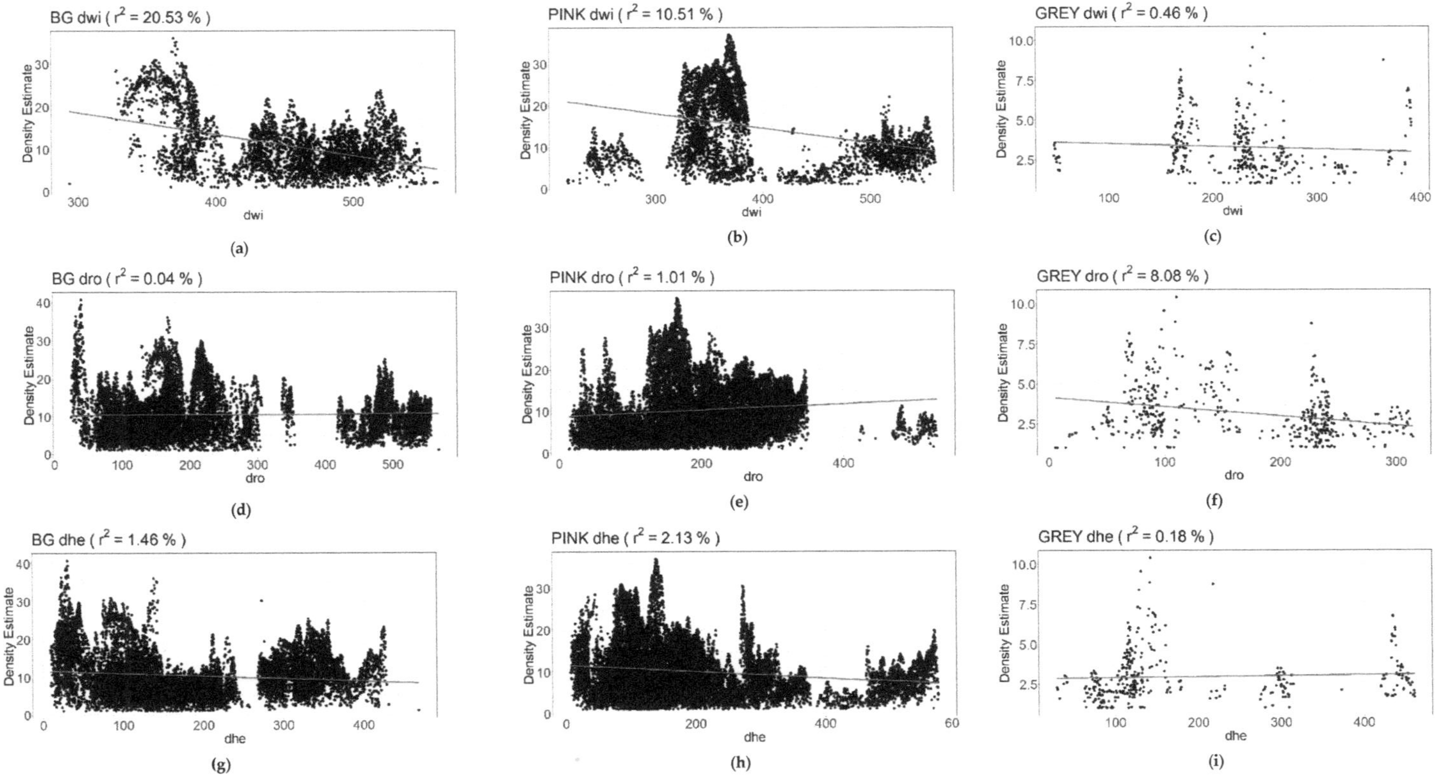

Figure 2. Linear regression of density (LRD) (illustrated as a red line), which regresses the density estimates of individual birds as a function of bird distances to the nearest landscape element. The following species were regressed: barnacle goose (BG), pink-footed goose (PINK), and Greylag goose (GREY). Distances from the landscape elements: wind turbines (dwi), roads (dro), and hedgerows (dhe). The r^2 values of LRD above 5% ($r^2 > 5\%$) are in bold.

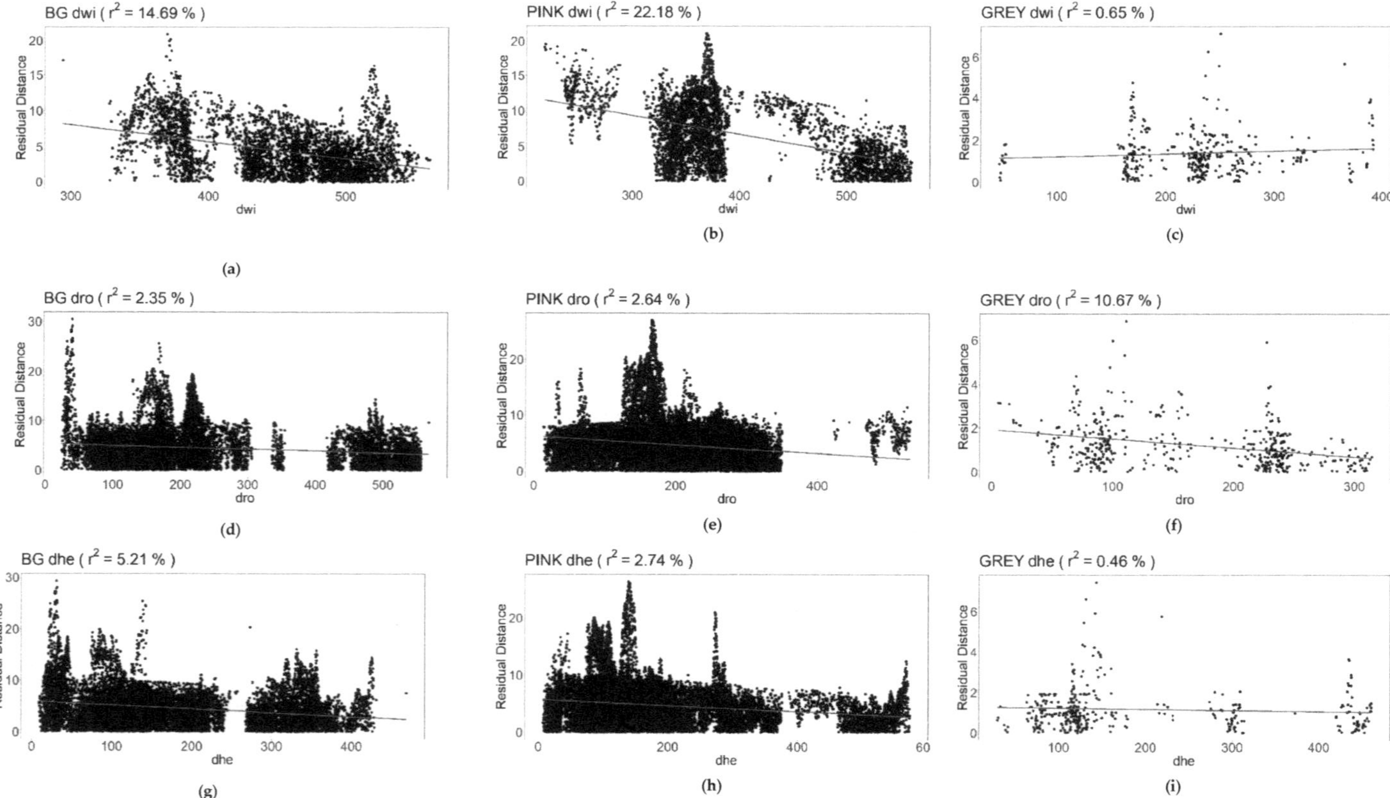

Figure 3. Linear regression of residuals (LRR) (illustrated as a red line), which regresses the absolute values of the residuals from LRD were regressed as a function of bird distances to the nearest landscape element. The following species were regressed: barnacle goose (BG), pink-footed goose (PINK), and greylag goose (GREY). Distances from the landscape elements; wind turbines (dwi), roads (dro), and hedgerows (dhe). The r² values of LRR above 5% (r² > 5%) are in bold.

3.2.2. Roads

The greylag geese showed a significant increase of BVAR with decreasing distance to the roads (r^2 = 10.67%) *** (Table 1 and Figure 3h). Whereas, barnacle geese and pink-footed geese showed a significant increase (with r^2 < 5%) increase of BVAR with decreasing distance to the roads (r^2 = 2.35%) *** and (r^2 = 2.64%) *** respectively (Table 1 and Figure 3b,e).

3.2.3. Hedgerows

The barnacle geese showed a significant increase of BVAR with decreasing distance to the hedgerows (r^2 = 5.20%) ***, (Table 1 and Figure 3c). Whereas, pink-footed geese showed a significant (with r^2 < 5%) increase of BVAR with decreasing distance to the hedgerows (r^2 = 2.74%) ***, (Table 1 and Figure 3f).

Lastly, the greylag geese showed (with a non-significant regression and r^2 < 5%) an increase of BVAR with decreasing distance to the hedgerows (r^2 = 0.46%, p > 0.05) ***, (Table 1 and Figure 3f).

4. Discussion

We have shown that the spatial distribution of foraging geese changes when they get closer to a disturbing landscape element (Table 1, Figures 2 and 3) and clear tendencies were observed indicating an increasing anti-predator cohesion of flocks of geese with decreasing distance to disturbing landscape elements. The change in spatial distribution is visible both for the behavioural instability of symmetry (BSYM) and behavioural instability of variance (BVAR), which will be discussed below.

There are clear tendencies, with few exceptions (with r^2 < 5%), for negative slopes of the linear regression of density (LRD), which indicates an increasing BSYM with decreasing distance to the landscape elements (Table 1 and Figure 2). There is also a clear tendency with few exceptions (with r^2 < 5%) for a negative slope of the linear regression of residuals (LRRs), which means an increasing BVAR with decreasing distance to the landscape elements (Table 1 and Figure 3). Thus, both indices show asymmetry of spatial distribution, implying environmental stress in flocks of geese induced by foraging near the studied landscape elements, which indicates these measurements to be useful tools for monitoring environmental stress. We have chosen the 5% threshold arbitrarily; however, it is also notable that the same negative trends were observed for other LRD and LRR although with r^2 below 5% (Figures 2 and 3). However, such relatively weak trends might not be of biological importance and should thus be interpreted with caution. Additionally, p-values also should be evaluated cautiously as the large sample size increases the significance of the indices even with low r2 as seen in both indices (Table 1). These relatively low r^2 values might be caused by noise from other factors influencing the density of bird flocks, masking the disturbing effects of the landscape elements. Such factors might include flock size [11] as well as food and water distribution [12], which are both known to affect the density and spatial distribution of bird flocks. Especially BVAR might have been affected by variations in flock size as flocks of different size prioritise anti-predation behaviour and foraging beneficial behaviour differently. A study by Lazarus [11], who examined the influence of flock size on the vigilance of white-fronted geese (*Anser albifrons*), noted that the percentage of vigilant birds in a flock would decrease steeply at flock sizes above 200–300 individuals [11]. Hence, larger flocks might prioritise foraging beneficial behaviour while smaller flocks prioritise anti-predator behaviour through the dilution effect [7,11,13,14]. However, trends were still observed for both BSYM and BVAR that prove that both methods can be applied to wildlife behaviour with multiple random influences, as long as the factor of interest is measured across a correlated variable.

In our study linear relationships were assumed and therefore only linear regression has been utilized. However, BSYM and BVAR could also be applied in studies with a non-linear relationship between the independent value and the dependent value. In this case, the asymmetry on the left and right side of mean can vary and might be estimated by the first derivative, which is equivalent to the slope observed for a certain value of the dependent value. Therefore, the BSYM index can be considered the equivalent of the absolute values of the difference between right and left value of a trait, which was referred to as fluctuating asymmetry (FA1) [15] in a study by Palmer and Strobeck [15]. Whereas, BVAR indices can be considered the equivalent of the phenotypic variability of trait (Vp). For both indices, FA1 and Vp are often utilised as estimators of developmental instability (DI) [16].

5. Conclusions

We have tested both indices against distance to landscape elements as an independent variable; however, the indices could also be utilised for other possible variables, e.g., regression against time or temperature. The merit of BSYM and BVAR is that they do not need bilateral traits for the estimation of the instability, which in our case we have called behavioural instability as we are estimating a deviation from expected behaviour, which is a constant distance between every single individual due to a uniform distribution of the individuals. These indices have proven effective tools for assessing the effect of environmental stress on expected behaviour. Thus, the term behavioural instability has been proven useful for data other than bilateral data as discussed in Pertoldi et al. [1]. We expect that the application of these two indices BSYM and BVAR could have several applications in applied ecology as both indices can be regressed against environmental gradients or temporal series.

Author Contributions: M.B.-H., R.M.K. and C.P. conceptualised the project and did the methodology. C.P. provided the resources, while M.B.-H., R.M.K., J.H.F.C. and J.B.G. did the investigation and data curation. M.B.-H., R.M.K., D.B., J.B.G. and C.P. did the formal analysis and visualization. M.B.-H., R.M.K. and C.P. took the lead in writing the manuscript. All authors provided critical feedback and helped shape the research, analysis and manuscript. M.B.-H. and R.M.K. were responsible for the project administration as well as, research activities and planning, who were supervised by D.B., B.L., L.F.J. and C.P.

Funding: Mads Bech-Hansen, Rune M. Kallehauge and Cino Pertoldi were financed by the Aalborg Zoo Conservation Foundation (AZCF: grant number 2018-4).

Acknowledgments: We would like to thank the Lead Guest Editor and the anonymous reviewers for invaluable suggestions and comments on the manuscript.

Conflicts of Interest: The authors declare no conflict of interest.

References

1. Pertoldi, C.; Bahrndorff, S.; Novicic, Z.K.; Rohde, P.D. The Novel Concept of "Behavioural Instability" and Its Potential Applications. *Symmetry* **2016**, *8*, 135. [CrossRef]
2. Rees, E.C. Impacts of wind farms on swans and geese: A review. *Wildfowl* **2012**, *62*, 37–72.
3. Madsen, J. Impact of Disturbance on Field Utilization of Pink-footed Geese in West Jutland, Denmark. *Biol. Conserv.* **1985**, *33*, 53–63. [CrossRef]
4. Larsen, J.K.; Madsen, J. Effects of wind turbines and other physical elements on field utilization by pink-footed geese (Anser brachyrhynchus): A landscape perspective. *Landsc. Ecol.* **2000**, *15*, 755–764. [CrossRef]
5. Chudzińska, M.E.; van Best, F.M.; Madsen, J.; Nabe-Nielsen, J. Using habitat selection theories to predict the spatiotemporal distribution of migratory birds during stopover—A case study of pink-footed geese Anser brachyrhynchus. *Oikos* **2015**, *124*, 854–860. [CrossRef]
6. Harrison, A.L.; Petkov, N.; Mitev, D.; Popgeorgiev, G.; Gove, B.; Hilton, G.M. Scale-dependent habitat selection by wintering geese: Implications for landscape management. *Biodivers. Conserv.* **2018**, *27*, 167–188. [CrossRef]
7. Hötker, H. Birds: Displacement. In *Wildlife and Wind Farms, Conflicts and Solutions; Volume 1 Onshore: Potential Effects*; Perrow, M.R., Ed.; Pelagic Publishing: Exeter, UK, 2017; pp. 119–154.
8. Roberts, G. Why individual vigilance declines as group size increases. *Anim. Behav.* **1996**, *51*, 1077–1086. [CrossRef]
9. Bech-Hansen, M.; Kallehauge, R.M.; Lauritzen, J.M.S.; Sørensen, M.H.; Pertoldi, C.; Bruhn, D.; Laubek, B.; Jensen, L.F. Evaluation of disturbance effect on geese and swans (Anserinae) caused by an approaching unmanned aerial vehicle. *Bird Conserv. Int..* under review.
10. Silverman, B. *Density Estimation for Statistics and Data Analysis*; Chapman & Hall: London, UK, 1986; pp. 9–13.

11. Lazarus, J. Vigilance, flock size and domain of danger size in the White-fronted Goose. *Wildfowl* **1978**, *29*, 135–145.
12. Gill, J.A. Habitat Choice in Pink-Footed Geese: Quantifying the Constraints Determining Winter Site Use. *J. Appl. Ecol.* **1996**, *33*, 884–892. [CrossRef]
13. Black, M.J.; Carbone, C.; Wells, R.L.; Owen, M. Foraging dynamics in goose flocks: The cost of living on the edge. *Anim. Behav.* **1992**, *44*, 41–50. [CrossRef]
14. Carbone, C.; Thompson, W.A.; Zadorina, L.; Rowcliffe, J.M. Competition, predation risk and patterns of flock expansion in barnacle geese (Branta leucopsis). *J. Zool. Lond.* **2003**, *259*, 301–308. [CrossRef]
15. Palmer, A.R.; Strobeck, C. Fluctuating asymmetry: Measurement, analysis, patterns. *Annu. Rev. Ecol. Syst.* **1986**, *17*, 391–421. [CrossRef]
16. Pertoldi, C.; Kristensen, T.N.; Andersen, D.H.; Loeschcke, V. Developmental instability as an estimator of genetic stress. *Heredity* **2006**, *96*, 122–127. [CrossRef] [PubMed]

Article

Evolutionary Game Research on Symmetry of Workers' Behavior in Coal Mine Enterprises

Kai Yu [1], Lujie Zhou [1,2,*], Qinggui Cao [1,*] and Zhen Li [1]

[1] College of Mining and Safety Engineering, Shandong University of Science and Technology, Qingdao 266590, China; ykxfkd@163.com (K.Y.); yskdk@sina.com (Z.L.)

[2] State Key Laboratory of Mining Disaster Prevention and Control Co-founded by Shandong Province and the Ministry of Science and Technology, Shandong University of Science and Technology, Qingdao 266590, China

* Correspondence: zhouqingdao@yeah.net (L.Z.); ykawyn@126.com (Q.C.); Tel.: +86-1396-989-1661 (L.Z.); +86-1396-984-9358 (Q.C.)

Received: 4 January 2019; Accepted: 28 January 2019; Published: 31 January 2019

Abstract: Statistics show that humans' unsafe behaviors are the main cause of accidents. Because of the asymmetry of game benefits between managers and coal miners, the stability of workers' behaviors is affected and unsafe behaviors are produced. In this paper, the symmetry of the behavior benefits of coal mine workers is studied, using game theory. In order to observe the dynamic game evolution process of behavioral stability, the paper establishes a system dynamics (SD) model and simulates it. The SD simulation results show that with the continuation of the game, when the benefits for safety managers and workers are asymmetric and the safety manager's safety inspection benefits are less than the non-inspection benefits, the manager may not conduct safety inspections, which poses a great hidden danger to safety production. Through dynamic incentives to regulate the symmetry of income of coal mine safety managers and coal mine workers, the purpose of enhancing the stability of safety behavior is achieved. The research results of the paper have been successfully applied to coal mine enterprises.

Keywords: benefits symmetry; behavioral stability; evolutionary game; system dynamics; dynamic incentive

---

1. Introduction

A large number of statistics show that in some large-scale accidental disasters, unsafe human behavior is the main cause of the accident [1–3]. More than 80% of coal mine accidents in China are directly or indirectly caused by unsafe human behaviors [4]. In safety production, the game income of coal mine managers and workers is asymmetric, which leads to unstable behavior on both sides. Therefore, it is very important to study the symmetry of game benefits and behavior stability between managers and workers in coal mine enterprises.

In terms of coal mine accidents, Chen et al. studied the development trend of coal mine accidents in China, and analyzed the human factors, indicating that the unsafe behavior of human beings is the main cause of the accidents [5]. Nouri assessed the safety of the Iranian Gas Company and concluded that workers' unsafe behavior is the main cause of accidents [6]. Ianole studied applied behavioral economics and its trends [7]. Experts have shown from a wide range of studies the importance of unsafe human behavior to safety production.

If safety behaviors are studied in depth, we must first understand their specific definition. In the late 1970s, Heinrich proposed that people's unsafe behavior and material insecurity led to accidents [8]. Based on Heinrich, Reason proposed a model of human factors in the 1990s and established an accident model, which was named "cheese". The theories presented that the potential risks in the system are deeply defended at all levels and ultimately lead to accidents [9].

Based on the definition of unsafe behavior, behavioral symmetry [10] and behavioral stability [11] were studied. In the construction industry, safety behaviors are studied from the aspects of safety cognition and safety climate. For example, Fang et al. established a cognitive model of unsafe behavior of construction workers [12]. Lyu et al. studied delay behavior from the aspects of safety climate and safety results [13]. Safety leadership has a great influence on the evolution of safety behaviors. For example, Beatriz et al. studied the influence of safety leadership on workers' safety behaviors in the light of working conditions [14], and Shen studied the transfer mechanism of safety behaviors [15].

In summary, most of the studies focus on the factors, formation mechanisms, and models of safety behaviors, while few focus on the dynamic evolution process of safety behaviors. Therefore, the paper uses dynamic evolutionary game theory to analyze the evolution process of safety behaviors.

Game theory has been widely used in many fields of scientific research. It discusses open research topics of importance to economics and the broader social sciences [16]. Game theory has been used to study the improvement of resources and environment. Zhao et al. analyzed environmental protection materials and environmental risks [17]. Using game theory, Feng et al. put forward the method of optimizing resource allocation [18]. In social science, Lu et al. established a multi-party evolutionary game model [19]. Sun et al. focused on the game between employer and worker mobility [20]. Wang et al. studied the interaction among government, enterprises, and workers [21].

Game theory has made outstanding contributions in social science and other research. Combining game theory with safety behaviors of coal miners provides a new research direction for studying the evolution process of them. If the game theory is used alone, the evolutionary results of safety behaviors in equilibrium state can be obtained, but the evolutionary state of safety behaviors in each time period cannot be observed better. Therefore, the paper combines game theory with system dynamics, gives full play to system dynamics, and deeply studies the changing state of workers' safety behaviors in each time period.

Professor Forrester (Jay W. Forrester) of the Massachusetts Institute of Technology (MIT) pioneered System Dynamics (SD) in 1956, which is combined qualitative analysis with quantitative analysis to study the functions of complex systems and the interaction of behaviors through model simulations [22,23]. In the construction industry, system dynamics is used to analyze the safety behaviors of workers, such as the feedback mechanism of workers' safety attitudes and safety behaviors [24], the influencing factors of workers' unsafe behaviors [25], and the influence of workers' work interference and family conflict on safety behaviors [26]. In enterprise management, system dynamics is used to simulate the relationship between safety investment and coal mine accidents [27], and optimize the risk management of chemical enterprises [28]. In other research fields, system dynamics is also used as a research tool to analyze the dependence between safety factors [29], and the safety psychological process of railway workers [30].

The behavior benefits matrix of the safety management department workers group is constructed to solve the problem of asymmetry of benefits and unstable behaviors between the safety management department and multiple workers groups. Based on the matrix, the evolutionary game analysis is carried out on the behavior stability of coal mine workers. The evolutionary game method is combined with system dynamics (SD), and Vensim is used to build the evolutionary SD model of safety behaviors. The dynamic evolution rules of the safety management department and the two coal mine workers groups are deeply analyzed through simulation by SD.

2. Evolutionary Game Analysis

Coal mining enterprises usually include frontline units and safety management departments. Workers are the main producers of behavior, which determine whether coal mining enterprises can operate safely. As an external factor, managers play a direct role in supervising the behaviors of workers, and will assume corresponding responsibilities.

Under the inspection of safety managers, workers may follow safety instructions to take actions, or may not comply with instructions to choose risky operations [31–33]. Workers are affected by safety managers, but also affect the decision-making behavior of the managers. Therefore, it can be considered that there is a game relationship between the inspection of the safety management department and the behavior choice of coal mine workers, but in the actual production of coal mine enterprises, the safety management department is facing more than a group of workers. Therefore, the paper analyses the game between a safety management department and two groups of coal mine workers.

The relevant assumptions of the game are as follows:

(1) Limited rational game group: coal mine workers Group 1, coal mine workers Group 2, coal mine safety management department. The strategies of coal mine workers groups include "safe behavior" and "unsafe behavior". Safety management departments have "inspection" and "no inspection" strategies.

(2) It is assumed that managers and coal miners are rational economic persons who determine their behavior based on cost-benefit analysis. Coal mine workers do not consciously abide by safety operating rules at all times. Similarly, managers do not always carry out safety checks, but once they do, they will strictly abide by the rules and regulations. That is to say, if workers' unsafe behaviors are found, they will be punished according to safety management regulations [34–36].

(3) Coal mine safety management departments and coal miners have limited rational characteristics. That is, there will inevitably be errors in logical reasoning or decision-making judgment and self-interest considerations. The logical reasoning or decision-making judgment is illustrated by examples. Managers would spend a lot of time checking Group 1 if they thought that Group 1 often had unsafe behavior, while Group 1 thought that the manager would not carry out safety checks immediately after checking, so the safety consciousness of Group 1 began to slacken. Consideration of self-interest is illustrated by examples. Group 2 believes that inspectors will inspect Group 1, according to the usual practice. Unsafe behaviors occur repeatedly in a short period of time, while members of groups with high safety awareness choose conformity behaviors under group pressure. From a group perspective, this may move the coal miners far away from the optimal benefit decision-making choice.

(4) The r is the reward for safety behaviors of coal mine workers. The d is the cost of taking safe actions, such as the extra physical and time required to take safety behaviors. If unsafe behavior is taken, there is no need to pay the corresponding cost. On the contrary, this part of the benefit will be obtained. The e is the additional benefits (such as psychological benefits, economic benefits, time benefits, etc.) for the coal miners who take unsafe behaviors. If the safety management department finds unsafe behaviors, it will take measures to correct unsafe behavior, such as education and training, and impose fines. The F is a penalty for unsafe behaviors (mainly economic penalties). The H is the loss that the group itself needs to bear, when the group has unsafe behaviors. The income obtained by the safety management department from inspecting the behaviors of coal miners mainly comes from the punishment (F) for the unsafe behaviors of coal miners. The C is the cost of safety inspection. The D is the loss sustained by the coal mine safety management department when the coal mine workers take unsafe behaviors. The probability of workers Group 1 adopting unsafe behaviors is x_1, and that of Group 2 is x_2. The probability of safety management inspection is a_1.

The benefits matrix of Group 1 and Group 2 is shown in Table 1.

Table 1. The benefits matrix of Group 1 and Group 2.

Benefits of Group 1	Benefits of Group 2	
	Unsafe Behaviors (x_2)	Safety Behaviors ($1 - x_2$)
Unsafe Behaviors (x_1)	$(d_1 - H_1 - F_1 - a_1 \times F_2, d_2 - H_2 - F_2 - a_1 \times F_1)$	$(d_1 - H_1 - F_1 - a_1 \times F_2, r_2 - d_2)$
Safety Behaviors ($1 - x_1$)	$(r_1 - d_1, d_2 - H_2 - F_2 - a_1 \times F_1)$	$(r_1 - d_1, r_2 - d_2)$

Group 1 chooses the benefits of safe behaviors v_{11} and the benefits of choosing unsafe behaviors v_{12} as Equations (1) and (2).

$$v_{11} = (r_1 - d_1) \times a_1 \times x_2 + (1 - a_1) \times x_2 \times (r_1 - d_1) + (r_1 - d_1) \times a_1 \times (1 - x_2) + (1 - a_1) \times (1 - x_2) \times (r_1 - d_1) \tag{1}$$

$$v_{12} = a_1 \times x_2 \times (d_1 - H_1 - F_1 - a_1 \times F_2) + a_1 \times (1 - x_2) \times (d_1 - H_1 - F_1 - a_1 \times F_2)$$
$$+ (1 - a_1) \times x_2 \times (d_1 - H_1 - F_1 - a_1 \times F_2) + (1 - a_1) \times (1 - x_2) \times (d_1 -- H_1 - F_1 - a_1 \times F_2) \tag{2}$$

Therefore, Group 1 chooses the average expected benefits of safe behaviors and unsafe behaviors as Equation (3).

$$v_1 = x_1 \times v_{11} + (1 - x_1) \times v_{12} \tag{3}$$

The dx_1/dt represents the rate of change of the proportion of safe behaviors of Group 1 with time, and the replication dynamics of Group 1's selected safe behaviors can be obtained.

$$G(x_1) = dx_1/dt = x_1 \times (v_{11} - v_1) = x_1 \times (1 - x_1) \times (v_{11} - v_{12}) \tag{4}$$

Group 2 chooses the benefits of safe behaviors v_{21} and the benefits of choosing unsafe behaviors v_{22} as Equations (5) and (6).

$$v_{21} = (r_2 - d_2) \times a_1 \times x_1 + (1 - a_1) \times x_1 \times (r_2 - d_2) + (r_2 - d_2) \times a_1 \times (1 - x_1) + (1 - a_1) \times (1 - x_1) \times (r_2 - d_2) \tag{5}$$

$$v_{22} = a_1 \times x_1 \times (d_2 - H_2 - F_2 - a_1 \times F_1) + a_1 \times (1 - x_1) \times (d_2 - H_2 - F_2 - a_1 \times F_1)$$
$$+ (1 - a_1) \times x_1 \times (d_2 - H_2 - F_2 - a_1 \times F_1) + (1 - a_1) \times (1 - x_1) \times (d_2 - H_2 - F_2 - a_1 \times F_1) \tag{6}$$

Therefore, Group 2 chooses the average expected benefits of safe behaviors and unsafe behaviors as Equation (7).

$$v_2 = x_2 \times v_{21} + (1 - x_2) \times v_{22} \tag{7}$$

The dx_2/dt represents the rate of change of the proportion of safe behaviors of Group 2 with time, and the replication dynamics of Group 2 selected safe behaviors can be obtained.

$$G(x_2) = dx_2/dt = x_2 \times (v_{21} - v_2) = x_2 \times (1 - x_2) \times (v_{21} - v_{22}) \tag{8}$$

The safety management department randomly competes with any of the two groups, and the benefits matrix is shown in Table 2.

Table 2. Benefits management matrix.

Strategy Selection for Groups 1 and 2	Safety Management Benefits	
	Safety Inspection (a_1)	Safety Non-Inspection ($1 - a_1$)
Both groups choose unsafe behaviors	$F_1 - D_1 + F_2 - D_2 - C$	$-D_1 - D_2$
Group 1 chooses Safety Behaviors; Group 2 chooses unsafe behaviors	$F_2 - D_2 - C$	$-D_1$
Group 2 chooses Safety Behaviors; Group 1 chooses unsafe behaviors	$F_1 - D_1 - C$	$-D_2$
Both groups choose safety behaviors	0	0

The benefits of safety managers' inspection and non-inspection are shown in Equations (9) and (10).

$$u_1 = x_2 \times (F_2 - D_2) + x_1 \times (F_1 - D_1) - C \tag{9}$$

$$u_2 = - x_2 \times D_2 - x_1 \times D_1 \tag{10}$$

Therefore, the average expected benefits of safety managers' inspection and non-inspection are shown in Equation (11).

$$u = a_1 \times u_1 + (1 - a_1) \times u_2 \tag{11}$$

The $\mathrm{d}u/\mathrm{d}t$ is used to represent the change rate of the proportion of safety inspection with time, and the replication dynamics of safety inspection can be obtained.

$$G(u) = \mathrm{d}u/\mathrm{d}t = a_1 \times (u_1 - u) = a_1 \times (1 - a_1) \times (u_1 - u_2) \tag{12}$$

3. Model Simulation Analysis

3.1. System Dynamics (SD) Simulation Model

Safety management departments in coal mine enterprises face more than one workers group. Based on game theory, the paper establishes a system dynamics model of game between managers and multiple workers groups, as shown in Figure 1.

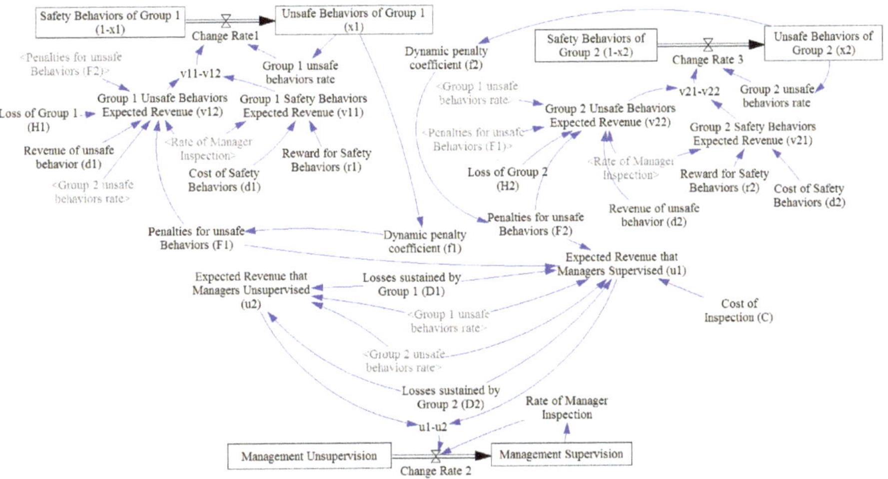

Figure 1. System dynamics (SD) model of evolutionary game.

The SD model of the evolutionary game has six state variables, three rate variables, 12 auxiliary variables, and 11 environmental variables. The system dynamics equation in the SD model is determined by the replication dynamic equation in the evolutionary game model.

3.2. Strategies Simulation Analysis

According to the above variables settings, the system dynamics model is analyzed by Vensim software (PLE, Ventana Systems. Inc, Harvard, MA., USA). The simulation results are shown in Figures 2–4.

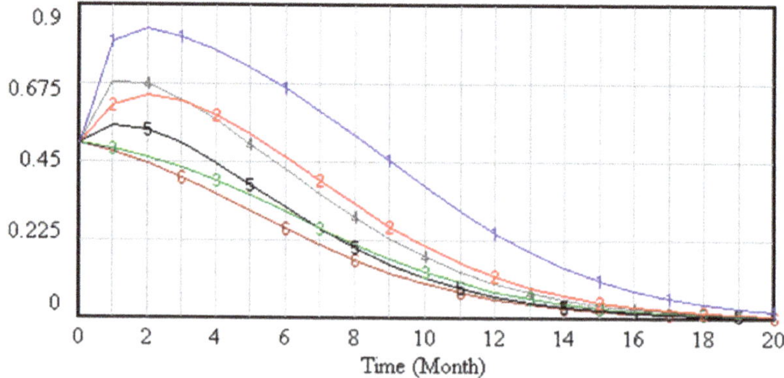

Figure 2. The probability of manager safety inspection under general state and dynamic penalty.

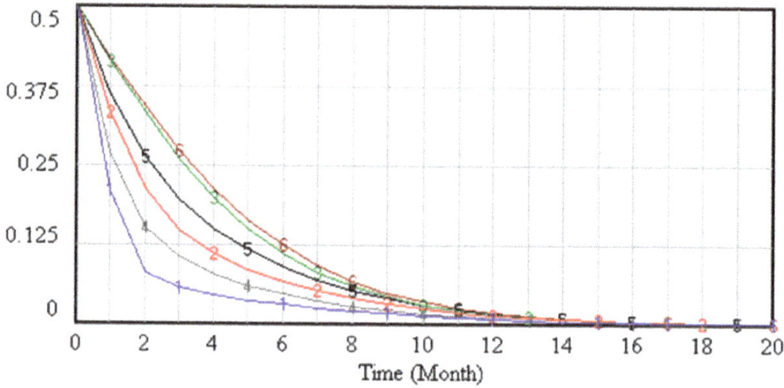

Figure 3. The probability of Group 1 choosing unsafe behaviors under general state and dynamic penalty.

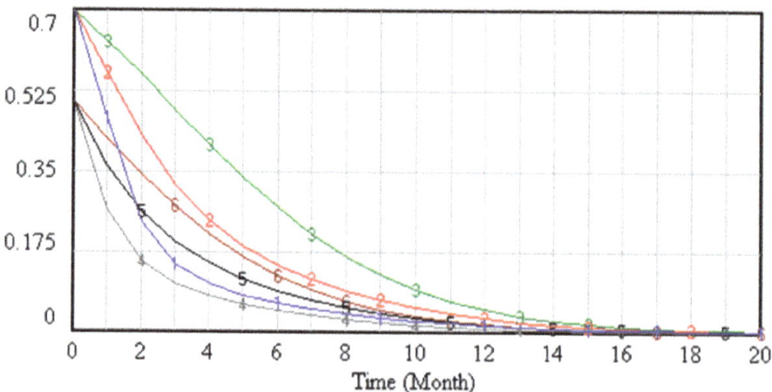

Figure 4. The probability of Group 2 choosing unsafe behaviors under general state and dynamic penalty.

In Figures 2–4, curves 1, 2, 4, and 5 show the dynamic evolution of strategies of each player when managers choose dynamic penalties. Curves 3 and 6 are the dynamic evolution of each player's strategy, when the probability of safety inspection is 50% and the unsafe actor is punished generally.

Curve 3 is the dynamic evolution of each player's strategy, when the level of unsafe behavior of both groups is 50%, the level of unsafe behavior of another group is 70%. Curve 6 is the dynamic evolution of the strategies of each player when the level of unsafe behaviors of two groups is 50%. Contrast curves 3 and 6 show that the level of unsafe behaviors of coal miners has an effect on the safety inspection of managers, while the level of unsafe behaviors of group 2 has little effect on the level of unsafe behaviors of group 1. Curves 3 and 6 in Figure 2 show that with the continuation of the game, asymmetry of the benefits between safety managers and workers began to appear. The level of group unsafe behaviors decreases, and the source of benefits for safety managers is 0. The safety manager chooses the safety probability not to rise but to fall. This will cause a great potential safety hazard to coal mining enterprises, which is not in line with the actual situation.

Curves 1 and 4, 2 and 5, and 3 and 6 show that when the punishment is positively correlated with the level of unsafe behavior, the greater the correlation coefficient is, the faster the level of unsafe behavior decreases, and the shorter the time it takes for the final group to choose safe behaviors. When the level of unsafe behaviors of two groups of coal miners is different, the two sides also influence each other, and the effect is obvious. This is because the level of safety behaviors of both sides has an impact on the behaviors of safety managers, and the punishment of the groups is also related to the level of unsafe behaviors of the other group.

Part of the benefit from coal mine management comes from fines for unsafe behaviors. Therefore, in order to maximize profits, safety management departments tend to choose safety inspection as their initial strategy. The higher the level of group unsafe behavior is, the more likely managers will choose safety management. With the improvement of the level of group safety behaviors, the benefit of the coal mine safety management department decreases. When the benefit of inspection is less than that of non-inspection, the probability of inspection decreases and finally evolves into a hidden danger state.

Therefore, only considering dynamic punishment can reduce the level of group unsafe behavior to a certain extent, but it can't motivate safety managers to carry out safety inspection continuously. This should consider how to adjust the benefit symmetry of managers and workers. In order to stimulate safety managers to conduct continuous safety inspections, dynamic incentives are considered here. The magnitude of dynamic incentive coefficient is negatively correlated with the level of group unsafe behaviors, that is, the greater the level of group safe behaviors, the greater the incentive of coal mine enterprises to the coal mine safety management department, and the greater the incentive to the group. The SD model with dynamic incentives is shown in Figure 5, and the simulation results are shown in Figures 6–8.

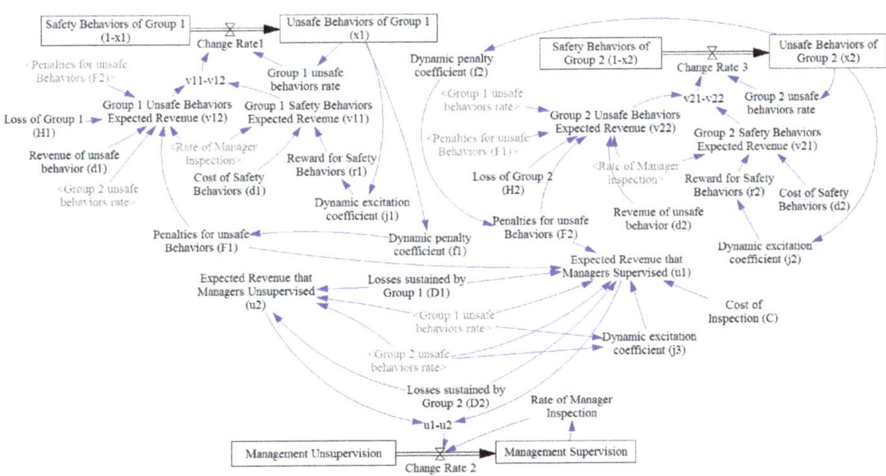

Figure 5. SD model with dynamic incentives.

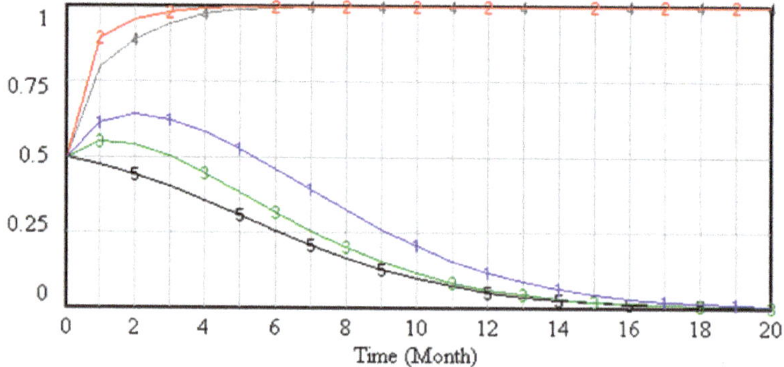

Figure 6. The probability of managers choosing safety inspection with dynamic incentives.

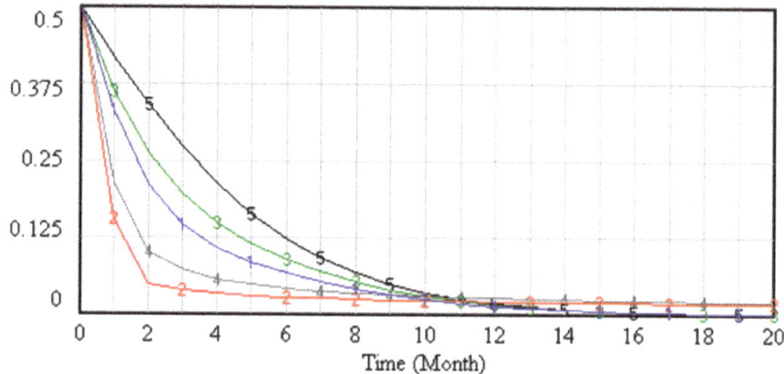

Figure 7. The probability of Group 1 choosing unsafe behaviors with dynamic incentives.

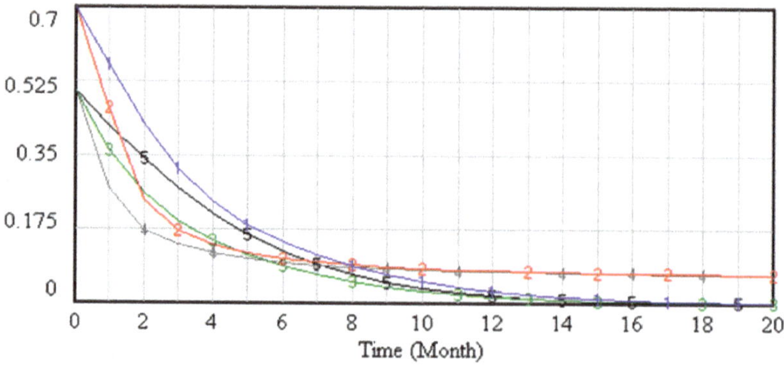

Figure 8. The probability of Group 2 choosing unsafe behaviors with dynamic incentives.

From the curves 2 and 4 of Figure 6, it can be seen that introducing dynamic incentives can improve the symmetry of game benefits between managers and workers. At the same time, it can also motivate safety managers to carry out safety inspections continuously and ensure the stability of the safe operation of coal mine workers. It can also be seen that when the level of unsafe behavior of two groups is not the same, the lower the level of unsafe behavior of the groups, the more quickly safety managers can make a safety inspection.

It can be seen from Figures 7 and 8 that after with the dynamic incentives, part of the benefits of coal mine workers depend on the level of safety behaviors of the groups. Therefore, the order of the decline rate of unsafe behavior level is: Dynamic incentive > Dynamic punishment > General punishment.

It is concluded from the simulation analysis that the dynamic incentives have a good effect on improving the benefit symmetry and behavior stability of managers and workers, and can promote the continuous supervision of safety managers and reduce the level of unsafe behavior of coal miners. Therefore, in the process of safety production, coal mine enterprises should give reasonable rewards and punishments to safety managers and workers to regulate the benefit symmetry of managers and workers and encourage them to choose safety behaviors, so as to ensure the smooth progress of safety production.

4. Example Application

Through the game analysis and SD simulation analysis of the safety management department and workers groups, it can be seen that dynamic incentives play a positive role in the stability of safety behaviors. Under the guidance of the theory, the Tangkou Coal Industry has been investigated on the spot. The existing violation treatment system has been optimized, the main content of which is to introduce a dynamic incentive system. Figure 9 shows the trend of violations of regulations in the Machinery and Electricity Team of Tangkou Coal Industry from January 2017 to November 2018. The team began to implement the dynamic incentive system in November 2017.

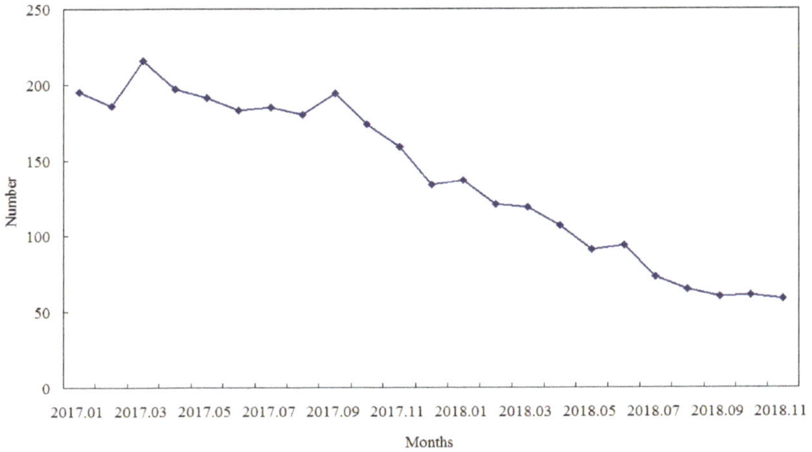

Figure 9. The trend of violations of regulations in the Machinery and Electricity Team from January 2017 to November 2018.

As can be seen from Figure 9, before the implementation of the dynamic incentive system (January 2017–October 2017), the number of violations of the team has been between 170–220. However, starting from November 2017, after the implementation of the dynamic incentive system, the number of violations of the team dropped rapidly. The number of violations dropped to below 70 in August 2018. The number of violations was stable at around 60 in September 2018, October 2018, and November 2018.

This paper compares and analyzes the violations of the team in the same month in 2017 and 2018, in order to more clearly analyze the changes in violations before and after the implementation of the dynamic incentive system, as shown in Figure 10. It can be clearly seen from Figure 10 that after the dynamic incentive measures were adopted, the number of violations of the team decreased significantly during the same period.

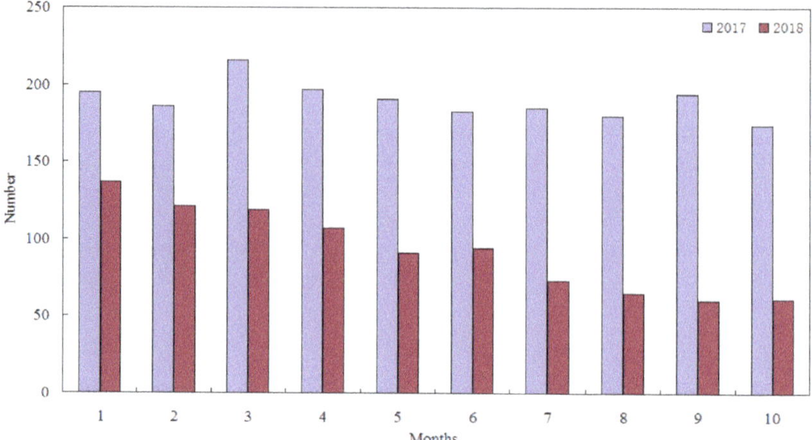

Figure 10. Comparison of the violations of the team in 2017 and 2018.

This shows that after the dynamic incentive measures, the safety management departments are motivated to be more active in the regulation of violations. At the same time, under the dynamic incentive measures, workers can also get extra benefits from safety production in the process of work, and their safety awareness is also improved.

It can be seen from Figures 9 and 10 that although the number of violations of the team has been significantly reduced, it is still around 60. This shows that there is still room for decline in the team's violations. Combined with the actual situation of Tangkou Coal Industry, if the dynamic incentive system can continue to be properly adjusted, the number of violations of the team may continue to decline. This is also the focus of the next research work in the paper.

5. Conclusions

The paper studies the symmetry of game benefits and behavior stability between coal mine safety managers and two groups of coal mine workers. Through the SD model, the evolutionary game process between safety management department and two groups of coal miners is analyzed in detail. The main conclusions are as follows:

(1) The symmetry of game benefits between safety managers and workers is analyzed, and their benefit matrix and game model is established.

(2) On the basis of the benefit matrix and game model, the SD model is constructed. The evolutionary process of game between the safety management department and two workers groups is analyzed in detail. The research shows that general punishment and dynamic punishment have a certain effect on reducing the safety behavior of coal miners, but at the same time, they also have some hidden dangers. Because the main source of benefit of coal mine safety managers is the fines for unsafe behavior, when the benefit symmetry is broken and the benefits of the safety management department's safety inspection is less than that of non-inspection, the safety management department ultimately does not carry out the safety inspection, which can be a great hidden danger to coal mine safety production.

(3) The dynamic incentive system is added to the SD model. It is concluded from the simulation analysis of the SD model that dynamic incentives can effectively regulate benefit symmetry and behavior stability of managers and workers, and promote continuous inspection by safety managers and reduce the level of unsafe behavior of coal mine workers.

(4) The research results of this paper have been applied to coal mine enterprises. By introducing dynamic incentive system, the existing reward and punishment system of enterprises has been adjusted and optimized. It effectively reduces the number of violations in coal mine enterprises.

The simulation analysis and practical application in this paper show that the establishment of a reasonable dynamic incentive mechanism can effectively improve the safety level of coal mine enterprises in the process of safety production.

Author Contributions: Conceptualization, K.Y. and Z.L.; methodology, K.Y. and L.Z.; validation, K.Y., L.Z., and Q.C.; investigation, K.Y., L.Z., and Q.C.; data curation, L.Z.; writing—original draft preparation, K.Y.; writing—review and editing, L.Z. and Q.C.; visualization, K.Y. and Z.L.; supervision, L.Z. and Q.C.; project administration, L.Z. and Q.C.; funding acquisition, L.Z. and Q.C.

Funding: This research was funded by National Natural Science Foundation of China, grant number 51474138; 51574157; 51804180.

Acknowledgments: The authors would like to thank the National Natural Science Foundation of China and the authors of the references.

Conflicts of Interest: The authors declare no conflict of interest.

References

1. Choudhry, R.M.; Fang, D. Why operatives engage in unsafe work behavior: Investigating factors on construction sites. *Saf. Sci.* **2008**, *46*, 566–584. [CrossRef]
2. Abdelhamid, T.S.; Everett, J.G. Identifying root causes of construction accidents. *J. Constr. Eng. Manag.* **2000**, *126*, 52–60. [CrossRef]
3. Shappell, S.A.; Wiegmann, D.A. A human error approach to accident investigation: The taxonomy of unsafe operations. *Int. J. Aviat. Psychol.* **1997**, *7*, 269–291. [CrossRef]
4. China Administration of Work Safety. [EB/OL]. 2017. Available online: http://www.chinasafety.gov.cn/newpage/ (accessed on 11 March 2017).
5. Chen, H.; Qi, H.; Long, R.; Zhang, M. Research on 10-year tendency of China coal mine accidents and the characteristics of human factors. *Saf. Sci.* **2012**, *50*, 745–750. [CrossRef]
6. Nouri, J.; Azadeh, A.; Mohammad, F.I. The evaluation of safety behaviors in a gas treatment company in Iran. *J. Loss Prev. Process Ind.* **2008**, *21*, 319–325. [CrossRef]
7. Ianole, R. *Applied Behavioral Economics Research and Trends*; IGI Global: Hershey, PA, USA, 2017; pp. 1–255.
8. Heinrich, H.W. *Industrial Accident Prevention*; McGraw-Hill: New York, NY, USA, 1979; pp. 19–26.
9. Reason, J. A systems approach to organizational error. *Ergonomics* **1995**, *8*, 1708–1721. [CrossRef]
10. Ahmed, R.R.; Channar, Z.A.; Soomro, R.H.; Vveinhardt, J.; Streimikiene, D.; Parmar, V. Antecedents of Symmetry in Physicians' Prescription Behavior: Evidence from SEM-Based Multivariate Approach. *Symmetry* **2018**, *10*, 721. [CrossRef]
11. Pertoldi, C.; Bahrndorff, S.; Kurbalija, N.Z.; Duun, R.P. The Novel Concept of "Behavioural Instability" and Its Potential Applications. *Symmetry* **2016**, *8*, 135. [CrossRef]
12. Fang, D.; Zhao, C.; Zhang, M. A Cognitive Model of Construction Workers' Unsafe Behaviors. *J. Constr. Eng. Manag.* **2016**, *142*. [CrossRef]
13. Lyu, S.; Hon, C.K.; Chan, A.P.; Wong, F.K.; Javed, A.A. Relationships among Safety Climate, Safety Behavior, and Safety Outcomes for Ethnic Minority Construction Workers. *Int. J. Environ. Res. Public Health* **2018**, *15*, 484. [CrossRef]
14. Fernández-Muñiz, B.; Montes-Peón, J.M.; Vázquez-Ordás, C.J. The role of safety leadership and working conditions in safety performance in process industries. *J. Loss Prev. Process Ind.* **2017**, *50*, 403–415. [CrossRef]
15. Shen, Y.; Ju, C.; Koh, T.Y.; Rowlinson, S.; Bridge, A.J. The Impact of Transformational Leadership on Safety Climate and Individual Safety Behavior on Construction Sites. *Int. J. Environ. Res. Public Health* **2017**, *14*, 45. [CrossRef] [PubMed]
16. Newton, J. Evolutionary Game Theory: A Renaissance. *Games* **2018**, *9*, 31. [CrossRef]
17. Zhao, R.; Neighbour, G.; Han, J.; McGuire, M.; Deutz, P. Using game theory to describe strategy selection for environmental risk and carbon emissions reduction in the green supply chain. *J. Loss Prev. Process Ind.* **2012**, *25*, 927–936. [CrossRef]
18. Feng, Q.; Cai, H.; Chen, Z.; Zhao, X.; Chen, Y. Using game theory to optimize allocation of defensive resources to protect multiple chemical facilities in a city against terrorist attacks. *J. Loss Prev. Process Ind.* **2016**, *43*, 614–628. [CrossRef]

19. Lu, R.; Wang, X.; Yu, H. Multiparty Evolutionary Game Model in Coal Mine Safety Management and Its Application. *Complexity* **2018**, *2018*, 9620142. [CrossRef]

20. Sun, J.; Wang, X.; Shen, L. Research on the mobility behaviour of Chinese construction workers based on evolutionary gametheory. *Econ. Res.-Ekon. Istraz.* **2018**, *31*, 1–14.

21. Wang, Q.; Mei, Q.; Liu, S. Analysis of Managing Safety in Small Enterprises: Dual-Effects of Employee Prosocial Safety Behavior and Government Inspection. *Biomed Res. Int.* **2018**, *2018*, 6482507. [CrossRef]

22. Zhong, Y. *System Dynamics*; Science Press: Beijing, China, 2009; pp. 14–19.

23. Wang, Q. *System Dynamics*; Science Press: Beijing, China, 2009; pp. 25–33.

24. Shin, M.; Lee, H.-S.; Park, M. A system dynamics approach for modeling construction workers' safety attitudes and behaviors. *Accid. Anal. Prev.* **2014**, *68*, 95–105. [CrossRef]

25. Bouloiz, H.; Garbolino, E.; Tkiouat, M. A system dynamics model for behavioral analysis of safety conditions in a chemical storage unit. *Saf. Sci.* **2013**, *58*, 32–40. [CrossRef]

26. Wu, G.; Duan, K.; Zuo, J. System Dynamics Model and Simulation of Employee Work-Family Conflict in the Construction Industry. *Int. J. Environ. Res. Public Health* **2016**, *13*, 1059. [CrossRef] [PubMed]

27. Tong, L.; Dou, Y. Simulation study of coal mine safety investment based on system dynamics. *Int. J. Min. Sci. Technol.* **2014**, *24*, 201–205. [CrossRef]

28. Zhang, M.; Wang, X.; Mannan, M.S.; Qian, C.; Wang, J. A system dynamics model for risk perception of lay people in communication regarding risk of chemical incident. *J. Loss Prev. Process Ind.* **2017**, *50*, 101–111. [CrossRef]

29. Jiang, Z.; Fang, D.; Zhang, M. Understanding the Causation of Construction Workers' Unsafe Behaviors Based on System Dynamics Modeling. *J. Manag. Eng.* **2015**, *31*, 04014099. [CrossRef]

30. Da, X.; Zheng, W. A system dynamics model for railway workers' safety behaviors. In Proceedings of the 2016 IEEE International Conference on Intelligent Rail Transportation (ICIRT), Birmingham, UK, 23–25 August 2016; pp. 409–417.

31. Holovatch, Y.; Janke, W.; Thurner, S. Self-organization and collective behaviour in complex systems. *Condens. Matter Phys.* **2014**, *17*, 30001. [CrossRef]

32. Stephen, P.R. *Organizational Behavior*; Prentice Hall [Imprint]; Pearson Education, Limited: London, UK, 2010; pp. 17–32.

33. Hu, B. *The Principle and Application of Qualitative Simulation of Group Behavior*; Huazhong University of Science and Technology Press: Wuhan, China, 2006; pp. 27–39.

34. Yu, K.; Cao, Q.; Zhou, L. Study on qualitative simulation technology of group safety behaviors and the related software platform. *Comput. Ind. Eng.* **2019**, *127C*, 1037–1055. [CrossRef]

35. Cao, Q.; Yu, K.; Zhou, L.; Wang, L.; Li, C. In-depth research on qualitative simulation of coal miners' group safety behaviors. *Saf. Sci.* **2019**, *113C*, 210–232. [CrossRef]

36. Brocher, T. Orientation on group dynamics. *Psychiatr. Commun.* **1970**, *13*, 3–12.

Article

Route Choice Behavior: Understanding the Impact of Asymmetric Preference on Travelers' Decision Making

Kai Liu * and Yuan Xu *

School of Engineering and Management, Nanjing University, Nanjing 210093, China
* Correspondence: dg1415007@smail.nju.edu.cn (K.L.); xy_0011@163.com (Y.X.); Tel.: +86-159-520-35866 (K.L.); +86-180-129-68631 (Y.X.)

Received: 10 December 2018; Accepted: 4 January 2019; Published: 8 January 2019

Abstract: This paper investigated the impact of asymmetric preference on travelers' route choices. Firstly, a status quo-dependent route choice mode was developed to describe travelers' route choices. Then, based on that model, a route choice experiment was conducted, and during the experiment, participants were requested to choose a route from two arbitrary non-dominated routes. Finally, according to the observation data, data analysis and model parameter estimation were conducted. The results show that participants used different measures to trade off travel cost and travel time. Additionally, there was a gap between most participants' willingness to pay (WTP) and willingness to accept (WTA). Moreover, participants' WTP greater than their own WTA was the key reason resulting in the inertial route choices. The empirical results in this paper can help the traffic manager to understand travelers' inertial route choice behavior from a different perspective.

Keywords: asymmetric preference; inertial route choice; willingness to pay; willingness to accept; experimental study

---

1. Introduction

Travelers' route choice behaviors have been researched for many years. Traditionally, it was assumed that all travelers would choose the routes with the shortest travel time [1]. However, in reality, this assumption is very restrictive, because it ignores travelers' cognitive limitations and intrinsic preferences [2–4]. Moreover, many empirical studies have shown that travelers do not always choose the shortest routes [5–7]. Simon [8] indicates that people are boundedly rational, which means sometimes they prefer a satisfactory choice to an optimal one. Based on the concept of bounded rationality, Mahmassani and Chang [9] introduced the concept of the indifference band to study travelers' boundedly rational departure time choices. Lou et al. [10] further encapsulated the indifference band into route choice modeling and proposed a boundedly rational route choice model, which addressed the limitation on travelers' knowledge of traffic conditions and their capability of finding the best available routes. In their boundedly rational route choice model, travelers would not necessarily switch to the shortest route when the time (or cost) difference between the current route and the shortest one was lower than an inertia threshold. Zhao and Huang [11] applied the concept of aspiration level to investigate travelers' boundedly rational route choice behavior and found that travelers can only obtain the actual travel times of the chosen routes and are enabled to recognize the best routes.

The boundedly rational behavior in travel choice describes the facts that travelers will not always choose the routes with the shortest travel time. There are various causes raising such behavior, one of which is inertial choice. In fact, the term "inertia" was first mentioned in Newtonian physics, and it has been widely used to describe similar characteristics of human behavior. Inertial choice can be understood as the tendency to repeat previous choices [12]. In the domain of behavioral decision, inertia was always employed to characterize the tendency for decision makers to choose options that maintain the status quo regardless of its outcome in a subsequent decision scenario, though the concise definitions were slightly different among the literature [13–19].

In the area of travel behavior research, travelers' inertia was defined as the tendency to stick to the alternative that they had previously chosen [15,20–24]. Many empirical studies provided support for the inertial route choice behavior [7,25,26]. For example, Chorus and Dellaert [15] thought that the wish to save cognitive resources could lead travelers to inertial choices; Site and Filippi [27] explained that phenomenon in terms of "loss aversion"—the disadvantages of a move from the status quo are valued more heavily than the advantages. Besides saving cognition and avoiding loss aversion, other behavioral mechanisms which generate inertial behavior include the asymmetric preference [4,22], prevailing choice set [28], habitual behavior [29,30], familiarity, prior decision [21], risk aversion [31], endowment effect [32], and so on.

The asymmetric preference describes the phenomenon that travelers might feel different when they made a tradeoff between travel cost and travel time depending on whether they spend money to save time (willingness to pay, WTP) or they bear more time to receive money (willingness to accept, WTA) [33]. While in the conventional study of the value of time, the substitutions between money and time are assumed to be constant. Various theoretical and empirical studies have been developed to examine the value of WTP and WTA as well as to estimate the gap between those two values [34–38]. De Borger and Fosgerau [35] adopted the status quo as the reference state to distinguish travelers' judgment between gain and loss, and further examined the trade-off between money and time. Their empirical study confirmed the gap between WTP and WTA. However, De Borger and Fosgerau [35] used "wrong choice" to describe travelers' route choices in their research. In fact, travelers' route choices were not "wrong", their behavior was just coinciding with the inertial choice.

Therefore, based on the research of De Borger and Fosgerau [35], Xu et al. [4] proposed a status quo-dependent route choice model to handle the so called "wrong choice" behaviors from the view of inertial choice. In their proposed route choice model, travelers were assumed to "compare the travel cost to their status quo (travel cost of the currently used path) in deciding whether to switch to another alternative, and the underlying value of time is adaptive in the sense that it varies across different route choice contexts". Xu et al. [4] found that the status quo-dependent route choice model can explain the route choice inertia resulting from asymmetric preference. Moreover, the inertia is path-specific and can incorporate the scaling effect of travel cost on travelers' route choices.

However, there was a lack of empirical study in Xu et al.'s [4] research, thus they cannot clearly show that when travelers' WTP is smaller than their WTA, or when travelers' WTP is greater than their WTA, how travelers will make their route choices. In addition, how will the scaling effect of travel cost affect travelers' asymmetric preference? This paper is an extended work of Xu et al.'s research aiming to provide more detailed support as an empirical study. A route choice experiment was conducted to investigate the impact of asymmetric preference on travelers' inertial route choice. In the experiment, three different travel scenarios (short, middle, and long travel) were designed to collect participants' route choice data. Then, based on the experimental data, the value of participants' WTP and WTA were calculated and compared. According to the value of WTP and WTA, participants' route choice behaviors were analyzed. Moreover, the parameters in Xu's route choice model were also estimated.

The remainder of this paper is organized as follows. Section 2 describes the status quo-dependent route choice model on how to handle travelers' route choice behaviors that result from asymmetric preference. Section 3 describes the route choice experiment as well as the data collection procedure. Section 4 presents the calculation and analysis results of participants' WTP and WTA and route choices in the experiment. Section 5 presents the model parameters estimation. Finally, Section 6 concludes the paper.

2. The Status Quo-Dependent Route Choice Model

For the convenience of readers, the symbols used in this section are summarized as follows:

R_W	the set of feasible routes for a given origin-destination (OD) pair w
r	the index for a route $r \in R_w$
j	the index for a route $j \in R_w \backslash r$
t_r^w	the index for the travel time of route r between OD pair w
t_j^w	the index for the travel time of route j between OD pair w
τ_r^w	the index for the monetary cost of route r between OD pair w
τ_j^w	the index for the monetary cost of route j between OD pair w
c_r^w	the index for a pair of travel time and monetary cost of route r between OD pair w and $c_r^w = (t_r^w, \tau_r^w)$

Considering two arbitrary non-dominated routes r and j, $r, j \in R_w, w \in W$, without loss of generality, it is assumed that $t_r^w \leq t_j^w$ and $\tau_r^w \geq \tau_j^w$. For the travelers on route r, the status quo-dependent travel utility by changing to path j is calculated as Equation (1).

$$U\left(c_r^w, c_j^w\right) = -\lambda_1 \eta_1 \left(t_j^w - t_r^w\right) + \eta_2 \left(\tau_r^w - \tau_j^w\right) \tag{1}$$

where $U\left(c_r^w, c_j^w\right)$ is the status quo-dependent travel utility for the travelers on route r, with c_r^w representing the status quo and c_j^w the alternative. Parameter λ_1 is the travelers' loss aversion coefficient of the time cost. η_1 and η_2 are the coefficients that unify time and monetary factors with travel utility. Thus, travelers on route r would suffer a smaller travel time, but they would obtain a larger cost in the aspect of monetary factors, and travelers would stick to route r if and only if the inequality $U\left(c_r^w, c_j^w\right) \leq 0$ holds. Otherwise, the travelers on route r will change to route j. In addition, $U\left(c_r^w, c_j^w\right) \leq 0$ holds if and only if

$$-\lambda_1 \eta_1 \left(t_j^w - t_r^w\right) + \eta_2 \left(\tau_r^w - \tau_j^w\right) \leq 0 \Rightarrow \tau_r^w - \tau_j^w \leq \frac{\lambda_1 \eta_1}{\eta_2} \left(t_j^w - t_r^w\right). \tag{2}$$

As indicated by Equations (1) and (2), travelers on route r would not change to route j if the potential improvement of monetary cost is not great enough (i.e., less than or equal to the right-hand-side value), and the coefficient $\lambda_1 \eta_1 / \eta_2$ in Equation (2) is equal to the travelers' willingness to accept (WTA).

On the other hand, for the travelers on route j the status quo-dependent travel utility by changing to route r, is calculated as Equation (3).

$$U\left(c_j^w, c_r^w\right) = \eta_1 \left(t_j^w - t_r^w\right) - \lambda_2 \eta_2 \left(\tau_r^w - \tau_j^w\right) \tag{3}$$

where $U\left(c_j^w, c_r^w\right)$ is the status quo-dependent travel utility for the travelers on route j, with c_j^w representing the status quo and c_r^w the alternative. Parameter λ_2 is the travelers' loss aversion coefficient of the monetary cost. Travelers on route j would suffer a smaller monetary cost, but they would get a larger cost in the aspect of travel time and travelers would stick to route j if and only if the inequality $U\left(c_j^w, c_r^w\right) \leq 0$ holds. Otherwise, travelers on route j will change to route r. In addition, $U\left(c_j^w, c_r^w\right) \leq 0$ holds if and only if

$$\eta_1 \left(t_j^w - t_r^w\right) - \lambda_2 \eta_2 \left(\tau_r^w - \tau_j^w\right) \leq 0 \Rightarrow \tau_r^w - \tau_j^w \geq \frac{\eta_1}{\lambda_2 \eta_2} \left(t_j^w - t_r^w\right). \tag{4}$$

As indicated by Equations (3) and (4), travelers on route j would not change to route r if the increase in monetary cost is greater than or equal to the right-hand-side value, and the coefficient

$\eta_1 / \lambda_2 \eta_2$ in Equation (4) is equal to the travelers' willingness to pay (WTP). Then, based on the above analysis, a route choice experiment will be conducted in the following section to verify travelers' inertial route choices that resulted from asymmetric preference.

3. Experiment Design

The experiment was carried out in the laboratory of the School of Engineering and Management, Nanjing University, in November 2016. We used the z-Tree as our experimental platform [39] to deal with the inputs of all participants' route choices. Next, we will introduce in detail the participants, experimental design, and procedure in the experiment.

3.1. Participants

In total, 30 participants—12 females and 18 males—were recruited from Nanjing University, and all of them had never participated in a similar experiment. Participants were required to make route choice decisions for a total of 120 test rounds, according to the experimental settings and procedures. During the experiment, participants were not allowed to have any interaction with others.

3.2. Experimental Design

The experimental design included three travel scenarios reflecting three different levels of travel time—short, middle, and long. In each of the scenarios, there were two non-dominated routes, e.g., Route 1 and Route 2, and the route's status quo is characterized by (t_r, τ_r). During the experiment, participants needed to choose one's routes as their travel route, and in each scenario, participants would make route choices for 40 test rounds. Since we focused on travelers' route choice behaviors in this study, thus, the interaction between travelers and the collective effect of travelers' behavior on congestion are not considered in the experiment design. Tables 1 and 2 show the design of travel time and travel cost of two routes in the experiment.

Table 1. Investigating participants' willingness to pay (travel time *, monetary cost **).

Scenario 1 (Short)		Scenario 2 (Middle)		Scenario 3 (Long)	
Route 1	Route 2	Route 1	Route 2	Route 1	Route 2
$(25\,{}^{*}, 3\,{}^{**})$	$(10, \tau_2^1 + \alpha_n^1)$	$(50, 10)$	$(30, \tau_2^2 + \alpha_n^2)$	$(90, 20)$	$(60, \tau_2^3 + \alpha_n^3)$

* The unit of travel time is minute. ** The unit of monetary cost is Yuan, which is the Chinese currency unit.

Table 2. Investigating participants' willingness to accept (travel time, monetary cost).

Scenario 1 (Short)		Scenario 2 (Middle)		Scenario 3 (Long)	
Route 1	Route 2	Route 1	Route 2	Route 1	Route 2
$(24, \tau_1^1 + \beta_k^1)$	$(12, \hat{t}_2^1)$	$(45, \tau_1^2 + \beta_k^2)$	$(25, \hat{t}_2^2)$	$(85, \tau_1^3 + \beta_k^3)$	$(65, \hat{t}_2^3)$

Table 1 is used to investigate participants' WTP, and Table 2 is used to investigate participants' WTA. τ_1^i, τ_2^i, and $\hat{t}_2^i (i = 1, 2, 3)$ represent travel money costs in the three different travel scenarios, and to note that the values of τ_1^i, τ_2^i, and $\hat{t}_2^i$ are calculated based on the participants inputting answers about their WTP at the initial stage of the experiment. Figure 1 shows the experimental design steps. As shown in Figure 1, for a participant, at the initial stage of the experiment they needed to input their answers for "if the travel time can save 10 min, how much you willing to pay", "if the travel time can save 25 min, how much you willing to pay", and "if the travel time can save 45 min, how much you willing to pay". Then, based on the three inputted answers, we can calculate the values of τ_1^i, τ_2^i, and $\hat{t}_2^i (i = 1, 2, 3)$ in Tables 1 and 2. Additionally, from participants' own points of view, we can get the $(25, 3)$ is equal to $(10, \tau_2^1)$, $(24, \tau_1^1)$, and $(12, \hat{t}_2^1)$ (similarly, we can also get the $(50, 10)$ is equal to $(30, \tau_2^2)$, $(45, \tau_1^2)$, and $(25, \hat{t}_2^2)$; $(90, 20)$ is equal to $(60, \tau_2^3)$, $c_1 = (85, \tau_1^3)$, and $(65, \hat{t}_2^3))$. Note that in Tables 1 and 2 the parameters α_n^i and $\beta_k^i (i = 1, 2, 3)$ are constants, which are appended to the monetary cost, and the n and k index for the round of experiment. Moreover, as the experiment progressed, there were $\alpha_1^i > \alpha_2^i > \cdots > \alpha_t^i = 0 > \cdots > \alpha_T^i$ and $\beta_1^i > \beta_2^i > \cdots > \beta_t^i = 0 > \cdots > \beta_T^i$. This design ensures that participants will choose Route 1 in the inertial test rounds when investigating their WTP, and they will choose Route 2 in the inertial test rounds when investigating their WTA.

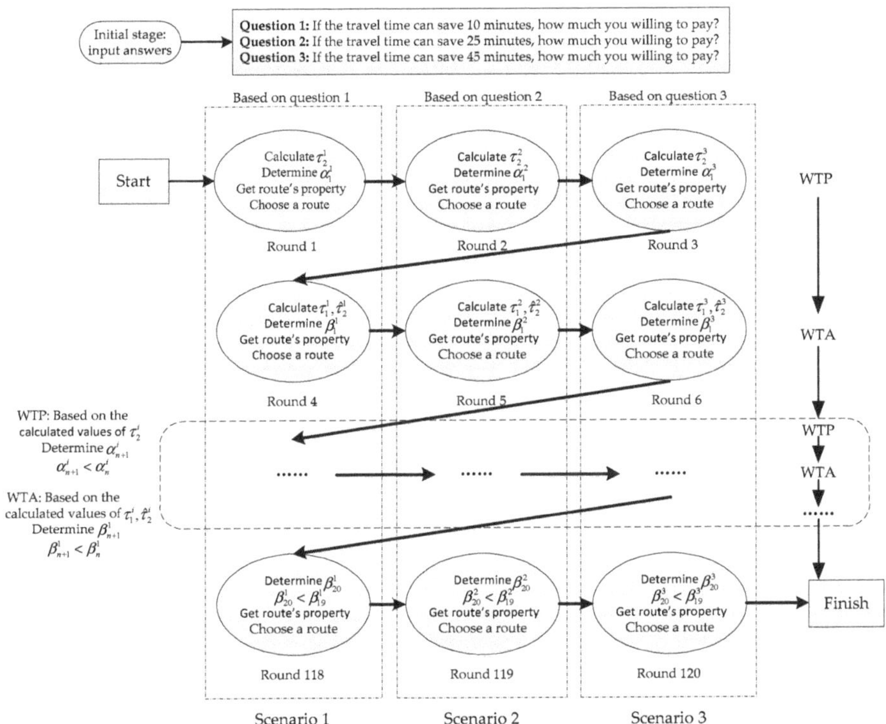

Figure 1. The design of experiment. WTP: willingness to pay; WTA: willingness to accept.

3.3. Procedure

Participants were invited to participate in the experiment at the University laboratory, and each participant was seated separately in front of a computer screen displaying the simulation. In order to encourage realistic behavior, participants were asked to imagine that they were commuters and driving to the workplace. The monthly salary was three times of their school living expenses. They were also informed that from the home to the workplace there were two routes that could be chosen. One route's travel time was larger than another route's travel time, but the travel money cost was smaller. For example, Route 1 was a congested road but toll-free, thus the monetary travel cost on Route 1 only consisted of an emission fee, and vehicle operating cost; Route 2 was a toll road but not congested, thus the monetary travel cost on Route 2 consisted of a congestion toll, emission fee, and vehicle operating cost. Note that in the cases of WTA, although a constant β_k^i was appended to the Route 1's monetary cost, Route 1 was still cheaper than Route 2 in the experiment. Figure 2 gives examples of actual choice situations in the experiment. No other information was provided before beginning the experiment.

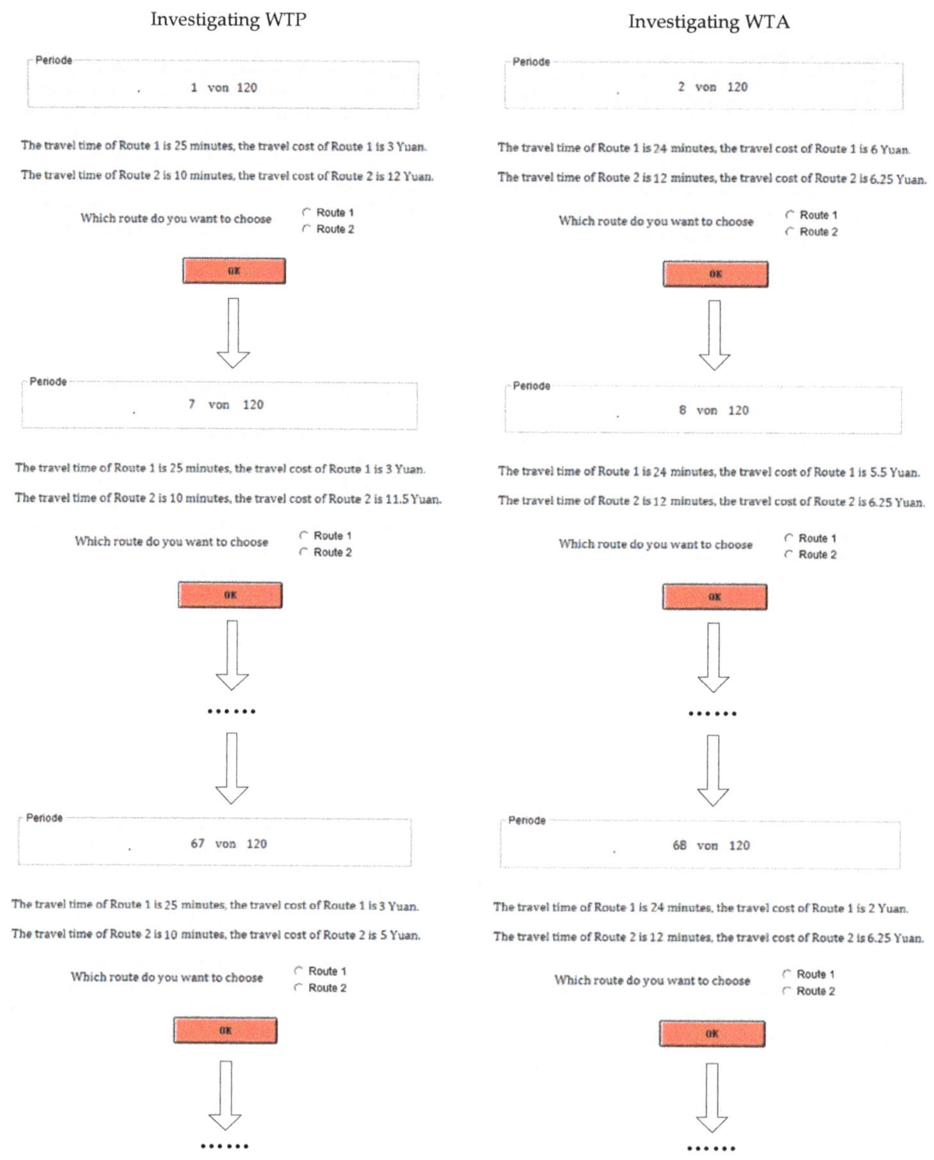

Figure 2. Snapshot of the route choice simulation window (Scenario 1).

The task included making a series of 120 consecutive route choices. Each test round represented a working day and participants were asked to make daily route choices while driving from home to workplace. Moreover, before the experiment, participants were requested to complete a post-task questionnaire that enquired about their loss aversion of the time cost and the monetary cost. The experimental steps are specified as follows:

(1) At the initial time, each participant needed to input the answers about their WTP, i.e., how much you are willing to pay if travel time can save 10 min; how much you are willing to pay if travel time can save 25 min; and how much you are willing to pay if travel time can save 45 min.

(2) During the experiment, in each test round, participants needed to choose one route from two non-dominated routes as their travel route, as can be seen in Figure 2. However, in order to reduce participants' arbitrary choices as made by the continuous choice under the same scenarios, the three choice scenarios alternated at the computer screen. Moreover, the interval time between each test round was one minute.

(3) The experiment ended, and each participant received a small gift as a show-up fee.

4. Results and Discussion

4.1. The Values of Participants' WTP and WTA

Based on the experimental data, the values of participants' WTP and WTA can be calculated. Note that the WTP refers to the value of a unit time saving in the case of loss in monetary cost, and WTA refers to the monetary compensation a traveler requires to agree to a unit travel time increase. Thus, in the experiment, the participants' WTP and WTA can be calculated as shown in Figure 3. It is worth mentioning that, for some participants, their initial stated WTP was different from the experimental WTP, in other words, this means that the result of the report is not equal to the result of the action. Considering the action result can better reflect participants' psychological index of decision-making, therefore, the experimental WTP and WTA were used in this paper. Figure 4 shows the calculation results. It clearly shows that for most participants, their WTP was smaller than WTA in all of three travel scenarios. This means that for these participants, the monetary compensation for the value of unit travel time increasing is larger than the monetary payment for the same value of unit travel time saving. Moreover, the test of null hypothesis shown in Table 3 suggests that the value of WTP is not equal to the value of WTA.

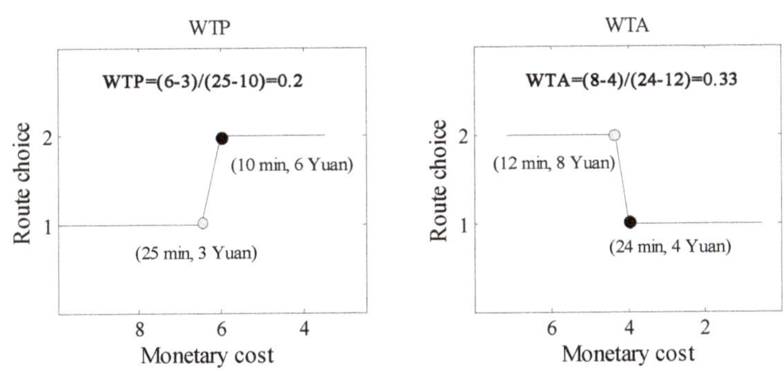

Figure 3. The calculation method of WTP and WTA.

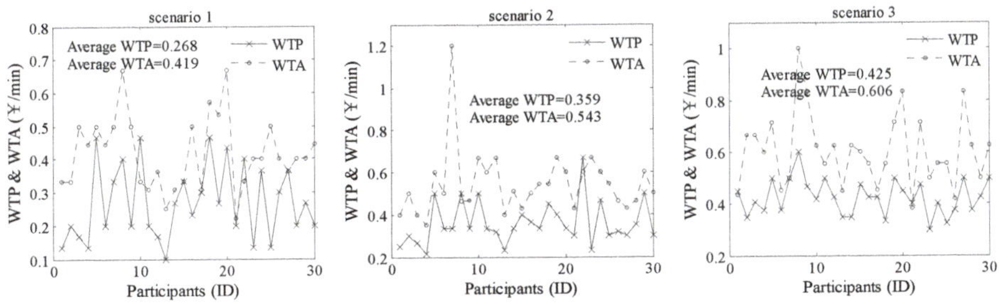

Figure 4. *Cont.*

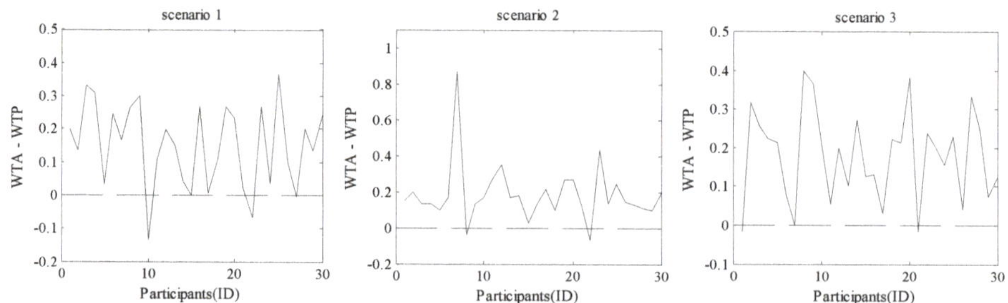

Figure 4. The value of WTP and WTA for all participants in three travel scenarios.

Table 3. Hypothesis test.

Null Hypothesis	*t*-Test	Std. err	95% Confidence Interval		Sig.
			Lower	Upper	
H_0: WTP = WTA	5.171	0.214	0.486	0.108	0.013

However, regardless of WTP < WTA or WTP > WTA, this result confirms that participants use different measures to distinguish the value of time, rather than constant substitution between money and travel time [33]. Thus, in the experiment, the asymmetric preference exists in participants' route choices. In addition, for the average values of the WTP, there is $WTP_1 < WTP_2 < WTP_3$ ($i = 1, 2, 3$ represents Scenario 1, Scenario 2, and Scenario 3). This result indicates that with the increase of the travel time, participants were willing to pay more to save travel time. Then, in order to explain the impact of asymmetric preference on participants' inertial route choices, in the following subsection, participants' route choice behaviors will be analyzed.

4.2. Influence of Asymmetric Preference on Participants' Route Choices

Since in the experiment, many participants' WTP < WTA or WTP > WTA, the impact of asymmetric preference on participants' route choices will be discussed with the two cases, i.e., WTP < WTA and WTP > WTA.

Case 1: WTP < WTA

In this subsection, Participant 2 will be taken as an example, and Figures 5–7 show Participant 2's route choices in the experiment under three different travel scenarios. As shown in Figure 5, in Scenario 1, Participant 2's WTP equals 0.2 and WTA equals 0.33. Then, assuming that if Participant 2's WTP equals WTA (i.e., WTP = WTA = 0.33), they should switch to Route 2 when its travel monetary cost is 8. However, in fact, Participant 2 continues to choose Route 1 until Route 2's travel monetary cost is 6. On the other hand, assuming that if Participant 2's WTA equals WTP (i.e., WTP = WTA = 0.2), they should switch to Route 1 when its travel monetary cost is 5.5. However, in fact, Participant 2 continues to choose Route 2 until Route 1's travel monetary cost is 4. Moreover, for Participant 2 (even those other participants whose WTP was smaller than WTA), similar phenomena can be found in the other two scenarios. Therefore, when the participants' WTP was smaller than WTA, the asymmetric preference can make them stick to the current routes until one alternative route's monetary cost is enough to tempt them to choose. This analysis result verifies that asymmetric preference (i.e., WTP < WTA) can lead travelers to make inertial choices (e.g., stick to the current route).

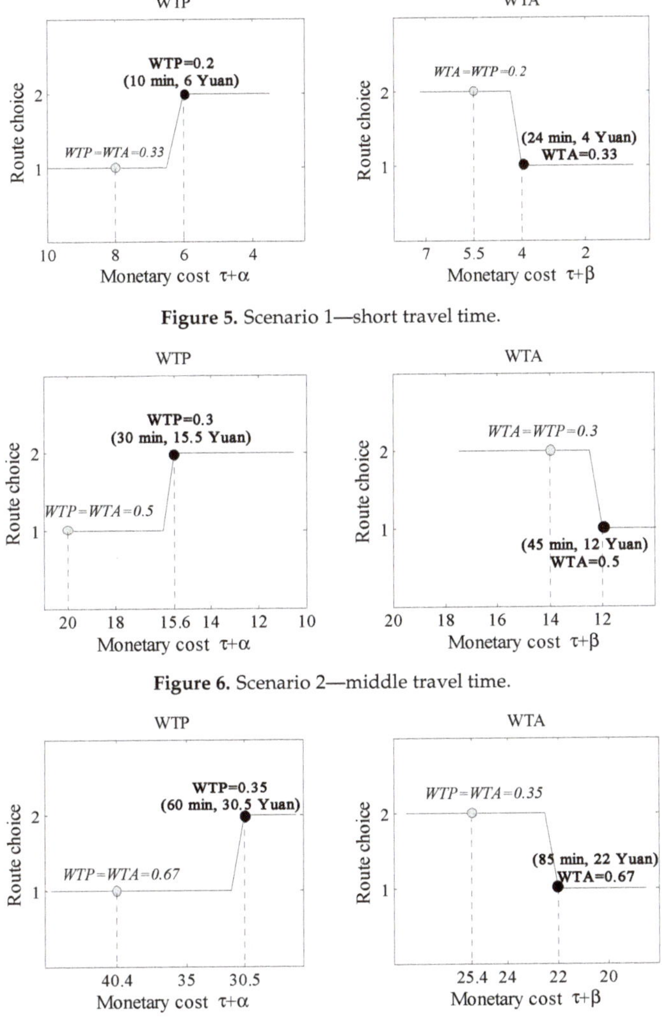

Figure 5. Scenario 1—short travel time.

Figure 6. Scenario 2—middle travel time.

Figure 7. Scenario 3—long travel time.

Case 2: WTP > WTA

In this subsection, Participant 10 (Scenario 1), Participant 22 (Scenario 2), and Participant 21 (Scenario 3) will be taken as examples, and Figures 8–10 show participants' route choices in the experiment under three different travel scenarios. As shown in Figure 8, in Scenario 1, Participant 10's WTP equals 0.47 and WTA equals 0.33. Then, assuming that if Participant 10's WTP equals WTA (i.e., WTP = WTA = 0.33), they should switch to Route 2 when its travel monetary cost is 8. However, in fact, Participant 10 switches to Route 2 when its travel monetary cost is 11. On the other hand, assuming that if Participant 10's WTA equals WTP (i.e., WTA = WTP = 0.47), they should switch to Route 1 when its travel monetary cost is 2.5. However, in fact Participant 10 switches to Route 1 when its travel monetary cost is 4.2. Moreover, for another two participants, similar phenomena can be found in the other two scenarios. Therefore, when participants' WTP was larger than WTA, the asymmetric preference can make them tend to choose a route with the shortest travel time.

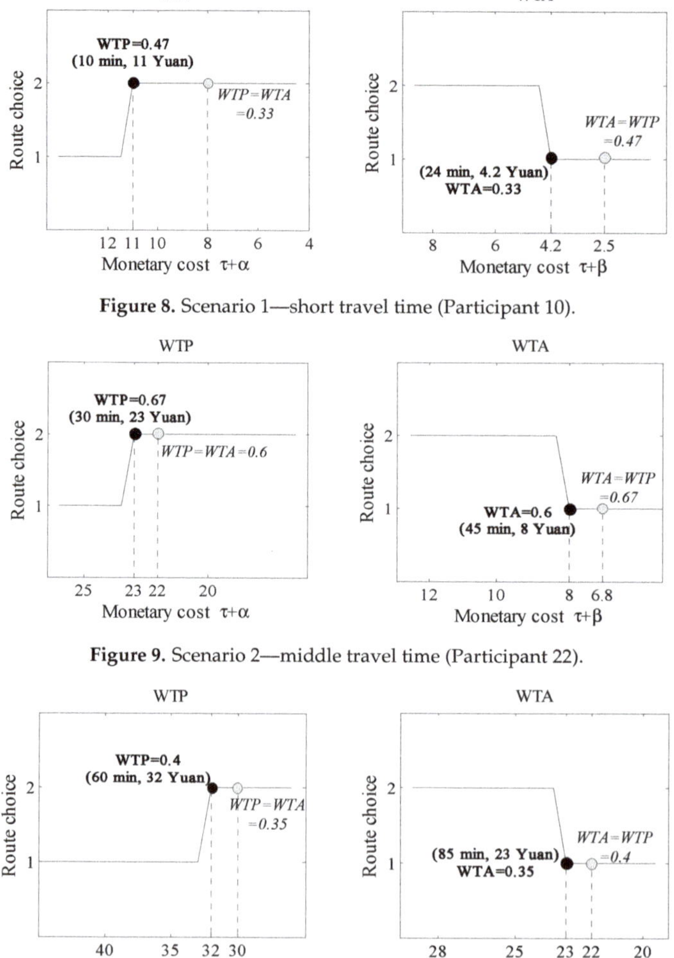

Figure 8. Scenario 1—short travel time (Participant 10).

Figure 9. Scenario 2—middle travel time (Participant 22).

Figure 10. Scenario 3—long travel time (Participant 21).

The above analysis results suggest that when participants' WTP is smaller than WTA, the asymmetric preference can make them stick to the current routes. Next, based on the observation data, the parameters λ_1, λ_2, η_1, and η_2 in Xu's [4] status quo-dependent route choice model will be estimated.

5. Model Parameters Estimation

5.1. Estimation of λ_1 and λ_2

The cumulative prospect theory (CPT), which was proposed by Tversky and Kahneman [40], is a valid utility measurement system. Many researchers used CPT to study individual's risk attitude and choice behavior under uncertainty over the past years. Xu et al. [41] developed a general travel decision-making rule utilizing CPT to investigate travelers' risk attitudes and route choice behaviors under uncertainty. Zhang et al. [42], based on the CPT, developed a day-to-day route-choice learning model with friends' travel information. Therefore, in this subsection, the cumulative prospect theory

(CPT) was applied to measure participants' risk attitudes, i.e., loss aversion of travel time and travel monetary cost. Without loss of generality, let x_0 define a reference point in the outcome domain, and the value function can be defined as Equation (5), and the weighting function can be defined as Equation (6).

$$v(x) = \begin{cases} (x - x_0)^\alpha, x > x_0 \\ -\lambda \cdot (x_0 - x)^\beta, x \le x_0 \end{cases} \tag{5}$$

$$w(p_i) = p_i^\gamma / \left[p_i^\gamma + (1 - p_i)^\gamma \right]^{1/\gamma} \tag{6}$$

where the parameters $\alpha \le 1$, $\beta \le 1$ measure the degree of the diminishing sensitivity to change in both directions from the reference point x_0, parameter $\lambda \ge 1$ captures the degree of loss aversion, parameters $0 < \gamma < 1$ reflect the level of distortion in the probability judgment. p_i is the probability of an outcome x_i. Figure 11 illustrates the value function and the weighting function. Suppose an alternative denoted by the pair $(x; p)$ is composed of $m + n + 1$ possible outcomes $x_{-m} < \cdots < x_0 < \cdots < x_n$ with probabilities $p_{-m}, \cdots, p_n$, respectively. Then, the cumulative decision weights are defined as follows:

$$\pi_i^+(p_i) = w^+(p_i + \cdots + p_n) - w^+(p_{i+1} + \cdots + p_n), 0 \le i \le n - 1 \tag{7}$$

$$\pi_{-j}^-(p_{-j}) = w^-(p_{-m} + \cdots + p_{-j}) - w^-(p_{-m} + \cdots + p_{-j-1}), 1 - m \le -j \le 0. \tag{8}$$

Accordingly, the CPT value of $(x; p)$ can be calculated by Equation (9).

$$U(x; p) = \sum_{i=0}^{n-1} v(x) \cdot \pi_i^+(p_i) + \sum_{j=1-m}^{0} v(x) \cdot \pi_{-j}^-(p_{-j}) \tag{9}$$

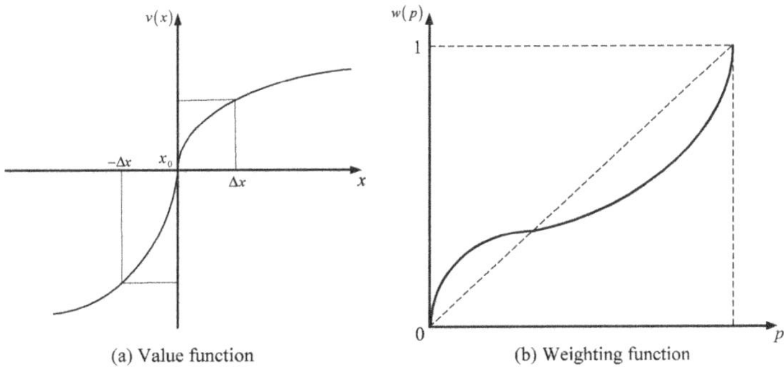

Figure 11. Value function and weighting function.

Firstly, participants were requested to complete a post-task questionnaire. Then, based on the survey data from the questionnaire, the parameter λ_1 and λ_2 were estimated. The design of the questionnaire considered the following two scenes:

Scene 1 (Investigate the loss aversion of the travel time): The travel times on both routes (Route A and Route B) are likely to increase because of congestion. Let $(t_1, p_1\%; t_2, p_2\%)$ denote the distribution of travel time on each route. For a traveler, there is a probability of $p_1\%$ to accomplish the trip with t_1 more minutes, a probability of $p_2\%$ to accomplish the trip with t_2 more minutes, and a probability of $1 - p_1\% - p_2\%$ with no extra time.

Scene 2 (Investigate the loss aversion of the travel monetary cost): The travel costs on both routes (Route R and Route S) are likely to increase because of the late penalty. Let $(c_1, p_1\%; c_2, p_2\%)$ denote the distribution of travel cost on each route. For a traveler, there is a probability of $p_1\%$ to accomplish the trip with c_1 more costs (Yuan, Chinese currency), a probability of $p_2\%$ to accomplish the trip with c_2 more costs, and a probability of $1 - p_1\% - p_2\%$ with no extra costs.

The route choice preferences of participants between A and B are shown in Table 4. The route choice preferences of participants between R and S are shown in Table 5.

Table 4. Results of route choice in Scene 1.

Case	Route A			Route B		
	Settings	**Number/Proportion**		**Settings**	**Number/Proportion**	
1	(20, 10%)	16	53.3%	(10, 20%)	14	46.7%
2	(20, 10%; 10, 80%)	20	66.7%	(10, 100%)	10	33.3%
3	(20, 10%; 10, 20%)	17	56.7%	(10, 40%)	13	43.3%
4	(20, 10%; 10, 70%)	14	46.7%	(10, 90%)	16	53.3%
5	(20, 10%; 10, 30%)	18	60.0%	(10, 50%)	12	40.0%

Table 5. Results of route choice in Scene 2.

Case	Route R			Route S		
	Settings	**Number/Proportion**		**Settings**	**Number/Proportion**	
6	(10, 10%)	17	56.7%	(5, 20%)	13	43.3%
7	(10, 10%; 5, 80%)	22	73.3%	(5, 100%)	8	26.7%
8	(10, 10%; 5, 20%)	16	53.3%	(5, 40%)	14	46.7%
9	(10, 10%; 5, 70%)	13	43.3%	(5, 90%)	17	56.7%
10	(10, 10%; 5, 30%)	19	63.3%	(5, 50%)	11	36.7%

Then, based on the survey data in Tables 4 and 5, the values of parameters λ_1 and λ_2 can be estimated using the least squares method and the cumulative squared residual can be given by following Equations (10) and (11), respectively:

$$f(\beta_1, \lambda_1, \delta) = \sum_{k=1}^{k=5} (A_k\% - P(A_k > B_k))^2 \tag{10}$$

$$f(\beta_2, \lambda_2, \delta) = \sum_{k=6}^{k=12} (R_k\% - P(R_k > S_k))^2 \tag{11}$$

where $A\%$ is the proportion of choosing Route A in case k, $P(A_k > B_k)$ is the probability that Route A is more valuable than Route B in case of k, and $P(A_k > B_k)$ can be given by Equation (12). $R\%$ is the proportion of choosing Route R in case of k, $P(R_k > S_k)$ is the probability that Route R is more valuable than Route S in case k, and $P(R_k > S_k)$ can be given by Equation (12). Here, $U(\cdot)$ is the route's cumulative prospect value and it can be calculated by Equation (9).

$$P(A_k > B_k) = \frac{1}{1 + \exp(U(B_k) - U(A_k))} \tag{12}$$

$$P(R_k > S_k) = \frac{1}{1 + \exp(U(S_k) - U(R_k))} \tag{13}$$

Note that in this paper, the value of parameter δ is set as 0.74, which was estimated by Wu and Gonzalez [43]. Participants' perception errors of the cumulative prospect value are assumed to follow the independent and identical Gumbel distributions. Then, based on these settings, the estimation values of β_1 is 0.35 and λ_1 is 1.41 in Scene 1; β_2 is 0.23 and λ_2 is 1.26 in Scene 2.

5.2. Estimation of η_1 and η_2

Until now, we have estimated the values of parameters λ_1 and λ_2, associated with the CPT-based utility measurement system. Specifically, $\lambda_1 = 1.41$ and $\lambda_2 = 1.26$. With them, based on the experimental data, we used the least squares method to estimate the values of parameters η_1 and η_2. Data of three thousand and six hundred observations of 30 participants were used. Table 6 presents the estimation results.

Table 6. Model estimation results.

	Scenario 1 (Short)		Scenario 2 (Middle)		Scenario 3 (Long)	
	η_1^1	η_2^1	η_1^2	η_2^2	η_1^3	η_2^3
Estimate	0.462	1.513	0.574	1.421	0.632	1.357
Std. error	0.214	0.476	0.246	0.432	0.284	0.407
t-Test	2.040	4.741	2.268	4.470	2.441	4.132
P-Value	<0.001	<0.001	<0.010	<0.001	<0.001	<0.001
$WTP = \eta_1/\lambda_2\eta_2$	0.242		0.321		0.364	
$WTA = \lambda_1\eta_1/\eta_2$	0.427		0.570		0.647	

As shown in Table 6, the estimation values of all parameters are positive and significant. Based on the estimation values of λ_1, λ_2, η_1, and η_2, participants' WTP (i.e., $\eta_1/\lambda_2\eta_2$) and WTA (i.e., $\lambda_1\eta_1/\eta_2$) can be calculated. Table 5 also shows the results of the calculation. As shown in Table 6, there is $WTP_1 < WTP_2 < WTP_3$ and $WTA_1 < WTA_2 < WTA_3$ ($i = 1, 2, 3$ represents Scenario 1, Scenario 2, and Scenario 3). These results are consistent with the aforementioned data analysis. Moreover, the calculation results of $WTP < WTA$ for three scenarios further explain that the asymmetric preference exists in participants' route choices.

6. Conclusions

In this research, a route choice experiment was conducted to investigate travelers' inertial route choice behaviors that resulted from asymmetric preference. Through data analysis, we can confirm that in the experiment participants used different measures to distinguish the value of time rather than constant substitution between money and travel time. Additionally, asymmetric preference significantly exists in participants' route choices. Especially, for most participants, their WTP was smaller than WTA, and this makes them stick to the current route until one alternative route's monetary cost is enough to tempt them to choose.

The value of this study lies in verifying travelers' inertial route choices that resulted from asymmetric preference and estimating the values of parameters in Xu's [4] status quo-dependent route choice model. Moreover, by considering the asymmetric preference, we can explain why some travelers stick to a toll lane, even when there is an alternative road that is slightly better. However, there are some limitations in our experimental design. Firstly, we used the questionnaire survey data of 30 participants to estimate the values of parameters λ_1 and λ_2. However, a behavioral experiment is needed to collect individuals' route-choice data to estimate the values of parameters λ_1 and λ_2. Secondly, as noted earlier, our experimental design was conducted with a lack of interaction between the participants' choices. Thus, there is meaningful reason for designing a behavioral experiment, and to incorporate the asymmetric preference into traffic equilibrium modeling.

Author Contributions: Model construction, experiment design, software, analysis of results and model parameters estimation: K.L.; introduction writing, model parameters estimation: Y.X.

Funding: This work was supported by a national natural science foundation of China [No. 71571097] and a humanities and social sciences project of Anhui, China [No. SK2018A0025].

Conflicts of Interest: The authors declare no conflict of interest.

References

1. Wardrop, J.G. Some theoretical aspects of road traffic research. *Proc. Inst. Civ. Eng.* **1952**, *1*, 325–378. [CrossRef]
2. Avineri, E. The effect of reference point on stochastic network equilibrium. *Transp. Sci.* **2006**, *40*, 409–420. [CrossRef]
3. Chen, R.B.; Mahmassani, H.S. Learning and risk attitudes in route choice dynamics. In *Expanding Sphere of Travel Behavior Research, Proceedings of the 11th International Conference, Orlando, FL, USA, 30 September–3 October 2009*; Kitamura, R., Yoshii, T., Eds.; Emerald: Bingley, UK, 2009; pp. 791–818.
4. Xu, H.L.; Yang, H.; Zhou, J.; Yin, Y.F. A route choice model with context-dependent value of time. *Transp. Sci.* **2017**, *51*, 536–548. [CrossRef]
5. Nakayama, S.; Kitamura, R.; Fujii, S. Drivers' route choice rules and network behavior: Do drivers become rational and homogeneous through learning? *Transp. Res. Rec.* **2001**, *1752*, 62–68. [CrossRef]
6. Avineri, E.; Prashker, J.N. Violations of expected utility theory in route-choice stated preferences: Certainty effect and inflation of small probabilities. *Transp. Res. Rec.* **2004**, *1894*, 222–229. [CrossRef]
7. Zhu, S. The roads taken: Theory and Evidence on Route Choice in the Wake of the I-35w Mississippi River Bridge Collapse and Reconstruction. Ph.D. Thesis, University of Minnesota, Mineapolis, MN, USA, 2010.
8. Simon, H.A. A Behavioral Model of Rational Choice. *Q. J. Econ.* **1955**, *69*, 99–118. [CrossRef]
9. Mahmassani, H.S.; Chang, G. On boundedly rational user equilibrium in transportation systems. *Transp. Sci.* **1987**, *21*, 89–99. [CrossRef]
10. Lou, Y.; Yin, Y.; Lawphongpanich, S. Robust congestion pricing under boundedly rational user equilibrium. *Transp. Res. Part B Methodol.* **2010**, *44*, 15–28. [CrossRef]
11. Zhao, C.L.; Huang, H.J. Experiment of boundedly rational route choice behavior and the model under satisficing rule. *Transp. Res. Part C Emerg. Technol.* **2016**, *68*, 22–37. [CrossRef]
12. Alós-Ferrer, C.; Hügelschäfer, S.; Li, J. Inertia and Decision Making. *Front. Psychol.* **2016**, *7*, 1–9. [CrossRef]
13. Huff, J.O.; Huff, A.S.; Thomas, H. Strategic renewal and the interaction of cumulative stress and inertia. *Strateg. Manag. J.* **1992**, *13*, 55–75. [CrossRef]
14. Sautua, S.I. Does Uncertainty Cause Inertia in Decision Making? An EXPERIMENTAL Study of the Role of Regret Aversion and Indecisiveness. Working Paper. 2016. Available online: https://www.researchgate.net/publication/301650273 (accessed on 8 January 2019).
15. Chorus, C.G.; Dellaert, B.G.C. Travel Choice Inertia: The Joint Role of Risk Aversion and Learning. *J. Transp. Econ. Policy* **2012**, *46*, 139–155.
16. Becker, G.S. *The Economic Approach to Human Behavior*; University of Chicago Press: Chicago, IL, USA, 1976.
17. Gal, D. A psychological law of inertia and the illusion of loss aversion. *Judgm. Decis. Mak.* **2006**, *1*, 23–32.
18. Kahneman, D.; Knetsch, J.L.; Thaler, R.H. The endowment effect, loss aversion, and status quo bias. *J. Econ. Perspect.* **1991**, *5*, 193–206. [CrossRef]
19. Miravete, E.J.; Palacios-Huerta, I. Consumer inertia, choice dependence, and learning from experience in a repeated decision problem. *Rev. Econ. Stat.* **2014**, *96*, 524–537. [CrossRef]
20. Chorus, C.G. Risk aversion, regret aversion and travel choice inertia: An experimental study. *Transp. Plan. Technol.* **2014**, *37*, 321–332. [CrossRef]
21. Van Exel, N.J.A. Behavioural Economic Perspectives on Inertia in Travel Decision Making. Ph.D. Thesis, Vrije Universiteit Amsterdam, Amsterdam, The Netherlands, 2011.
22. Lindsey, R. State-dependent congestion pricing with reference dependent preferences. *Transp. Res. Part B Methodol.* **2011**, *45*, 1501–1526. [CrossRef]
23. Train, K.E. *Discrete Choice Methods with Simulation*; The Cambridge University Press: Cambridge, UK, 2003.
24. Xie, C.; Liu, Z.G. On the stochastic network equilibrium with heterogeneous choice inertia. *Transp. Res. Part B Methodol.* **2014**, *66*, 90–109. [CrossRef]
25. Guo, X.; Liu, H.X. Bounded rationality and irreversible network change. *Transp. Res. Part B Methodol.* **2011**, *45*, 1606–1618. [CrossRef]
26. Vreeswijk, J.D.; Landman, R.L.; Van Berkum, E.C.; Hegyi, A. Improving the road network performance with dynamic route guidance by considering the indifference band of road users. *Intell. Transp. Syst.* **2015**, *9*, 897–906. [CrossRef]

27. Site, P.D.; Filippi, F. Stochastic user equilibrium and value-of-time analysis with reference-dependent route choice. *Eur. J. Transp. Infrastruct. Res.* **2011**, *2*, 194–218.
28. Zhang, J.L.; Yang, H. Modeling route choice inertia in network equilibrium with heterogeneous prevailing choice sets. *Transport. Res. Part. C Emerg. Technol.* **2015**, *57*, 42–54. [CrossRef]
29. Gärling, T.; Axhausen, K.W. Introduction: Habitual travel choice. *Transportation* **2003**, *30*, 1–11. [CrossRef]
30. Kim, H.; Oh, J.S.; Jayakrishnan, R. Effects of user equilibrium assumptions on network traffic pattern. *Ksce J. Civ. Eng.* **2009**, *13*, 117–127. [CrossRef]
31. Denrell, J.; March, J.G. Adaptation as information restriction: The hot stove effect. *Organ. Sci.* **2001**, *12*, 523–538. [CrossRef]
32. Thaler, R. Toward a positive theory of consumer choice. *J. Econ. Behav. Organ.* **1980**, *1*, 39–60. [CrossRef]
33. Small, K.A. Valuation of travel time. *Econ. Transp.* **2012**, *1*, 2–14. [CrossRef]
34. Boyce, R.R.; Brown, T.C.; McCledlland, G.H. An experimental examination of intrinsic values as a source of the WTS-WTP disparity. *Am. Econ. Rev.* **1992**, *82*, 1366–1373.
35. De Borger, B.; Fosgerau, M. The trade-off between money and travel time: A test of the theory of reference-dependent preferences. *J. Urban Econ.* **2008**, *64*, 101–115. [CrossRef]
36. Hess, S.; Rose, J.M.; Hensher, D.A. Asymmetric preference formation in willingness to pay estimates in discrete choice models. *Transp. Res. Part. E Logist. Transp. Rev.* **2008**, *44*, 847–863. [CrossRef]
37. Schmidt, U.; Traub, S. An experimental investigation of the disparity between WTA and WTP for lotteries. *Theory Decis.* **2009**, *66*, 229–262. [CrossRef]
38. Masiero, L.; Hensher, D.A. Shift of reference point and implications on behavioral reaction to gains and losses. *Transportation* **2011**, *38*, 249–271. [CrossRef]
39. Urs Fischbacher. z-Tree: Zurich toolbox for ready-made economic experiments. *Exp. Econ.* **2007**, *10*, 171–178. [CrossRef]
40. Tversky, A.; Kahneman, D. Advances in prospect theory: Cumulative representation of uncertainty. *J. Risk Uncertain.* **1992**, *5*, 297–323. [CrossRef]
41. Xu, H.L.; Zhou, J.; Xu, W. A decision-making rule for modeling travelers' route choice behavior based on cumulative prospect theory. *Transp. Res. Part. C Emerg. Technol.* **2011**, *19*, 218–228. [CrossRef]
42. Zhang, C.; Liu, T.L.; Huang, H.J.; Chen, J. A cumulative prospect theory approach to commuters' day-to-day route-choice modeling with friends' travel information. *Transp. Res. Part C Emerg. Technol.* **2018**, *86*, 527–548. [CrossRef]
43. Wu, G.; Gonzalez, R. Curvature of the probability weighting function. *Manag. Sci.* **1996**, *12*, 1676–1690. [CrossRef]

Article

Antecedents of Symmetry in Physicians' Prescription Behavior: Evidence from SEM-Based Multivariate Approach

Rizwan Raheem Ahmed [1,*], Zahid Ali Channar [2], Riaz Hussain Soomro [3], Jolita Vveinhardt [4], Dalia Streimikiene [5,*] and Vishnu Parmar [6]

[1] Faculty of Management Sciences, Indus University, Block-17, Gulshan, Karachi 75300, Pakistan
[2] Department of Business Administration, Sindh Madressatul Islam University, Karachi 74000, Pakistan; drzahidalic@gmail.com
[3] Institute of Health Management, Dow University of Health Sciences, Mission Road, Karachi 74200, Pakistan; riaz.soomro@duhs.edu.pk
[4] Faculty of Economics and Management, Vytautas Magnus University, Daukanto str. 28, 44248 Kaunas, Lithuania; Jolita.Vveinhardt@gmail.com
[5] Lithuanian Sports University, Institute of Sport Science and Innovations, Sporto str. 6, 44221 Kaunas, Lithuania
[6] Institute of Business Administration, University of Sindh, Jamshoro 76080, Pakistan; vishnu.parmar@usindh.edu.pk
* Correspondence: rizwanraheemahmed@gmail.com (R.R.A.); dalia.streimikiene@lei.lt (D.S.)

Received: 8 November 2018; Accepted: 3 December 2018; Published: 5 December 2018

Abstract: The aim of this paper is to examine the direct impact of marketing and medical tools on the symmetry of physicians' prescription behavior in the context of the Pakistani healthcare sector. This research also investigates the moderating influence of corporate image and customer relationship in an association of marketing & medical tools, and the symmetry of physicians' prescription behavior. The survey involved a research sample of 740 physicians, comprising 410 general practitioners and 330 specialists. A series of multivariate approaches such as exploratory factor analysis, confirmatory factor analyses, and conditional process analysis are employed. The findings of the study showed that marketing & medical tools have a direct, positive, and significant influence on physicians' symmetrical prescription behavior. Corporate image and customer relationship have also a significant impact as moderating variables between marketing & medical tools, and the symmetry of prescription behavior of physicians. The outcomes of this research are beneficial to marketers and medical managers in the pharmaceutical industry.

Keywords: pharmaceutical marketing; physicians' symmetrical prescription behavior; corporate image; exploratory factor analysis; confirmatory factor analysis; SEM-based multivariate approach

JEL Classification: C12; L6; M3

---

1. Introduction

Pharmaceutical marketing and medico marketing are different terminologies which are used in the pharmaceutical industry for the promotion of their brands and products in order to earn substantial revenues. The pharmaceutical companies use different marketing and medical strategies and symmetrical tactics such as medical journal advertising, medical detail aids, free drug samples, gifts, scientific activities (presentations in wards & doctors chambers, RTDs, DGMs, conferences, CMEs etc.), personal obligations, birthday greetings and gifts, airfare tickets, hotel accommodation, travel expenses, and recreational activities to the physicians and their families [1–4]. These promotional tools not only establish a good relationship with physicians but also bring augmented and symmetrical prescription business from physicians [3,5–7]. According to Ahmed et al. [1], and Masood et al. [8], the pharmaceutical industry is a lucrative business in Pakistan; therefore, there are several marketing and medical factors, which are helpful for generating the maximum revenue in the form of symmetrical prescription business. Thus, there is a strong need to re-evaluate the predictors (marketing & medical factors) that strongly influence physicians' symmetrical prescription behavior [3,9,10]. Modern marketing scholars and practitioners show ever-growing interest in the antecedents of symmetry in physicians' prescription behavior such as promotional material, medical activities, direct-to-consumer advertising, scientific activities, medical representatives' effectiveness, and other marketing & communications [1,11–13]. Similarly, pharmaceutical marketing costs have increased significantly over the last few years because of intense competition and deregulation of the pharmaceutical industry worldwide [14–16].

According to the IMS World Review 2017, pharmaceutical sales reached USD $1105 billion in the global market, indicating an increase of 8% in constant values compared to 2016 sales. The US pharmaceutical market's net worth is around USD341.1 billion, which is the largest in the world; it comprises approximately 36.5% of the global market. The North American market is lucrative due to effective healthcare systems, and the presence of major transnational pharmaceutical companies in the region. The Asia Pacific region pharmaceutical market is the second largest, comprising approximately 21.5% of the global market. However, the estimated size of the Pakistani pharmaceutical market is $3.40 billion, with average growth of 12% in 2017 [3]. During the same year, marketing expenditure of pharmaceutical products reached $6.1 billion, accounting for 6% growth as compared to the year 2016 for the global pharmaceutical industry. However, in Pakistan, electronic and print media advertising is allowed only for over-the-counter (OTC) drugs; therefore, there is an increasing tendency of marketing expenditure for prescription drugs in order to compete with rivals, especially due to the intensifying power of generic drugs. The major share of generic or copy brands accounts for more than 58.5% of the total pharmaceutical market in Pakistan [3].

Thus, given the above facts, there is tremendous pressure on pharmaceutical companies to justify marketing expenditures in terms of returns on investments, and measuring the marketing efforts that justify these investments. Since the pharmaceutical industry is quite unique, and intense regulations apply on the advertising of pharmaceutical drugs, this industry has adopted very aggressive personal promotional activities to indirect customers (physicians) that cost huge sums of money [1,17]. The business structure of the pharmaceutical industry is quite different to that of other industries. In the pharmaceutical industry, the consumers (Patients) are not the direct customers; rather, the physician is a decision maker on behalf of the patient. Thus, the physician is the direct customer of the pharmaceutical industry. The cost of the prescription is borne by the healthcare system or the patient himself; therefore, a physician is free to choose any medicine, whether it is expensive or cheap. The physician is in the driving seat, and he/she takes benefits from both sides, such as charging a fee from the patients/health insurance companies, while also receiving huge benefits from the pharmaceutical companies for prescribing their drugs [3,18]. The existing literature pertaining to pharmaceutical marketing has fixated on the effect of advertising cost on price elasticity, and the dissemination of new drugs, i.e., the influence of medical detailing (a form of personal selling), the role of marketing mix symmetry (samples, gifts, scientific activities, sponsorships etc.), the impact of customer relationships on the prescription business, the influence of direct-to-consumer advertising (DTCA), and personal obligations on physicians' symmetrical prescription behavior [1,19–21].

2. Significance and Objectives of the Study

However, pharmaceutical practitioners and research scholars have published numerous studies pertaining to the marketing strategies and their effectiveness in order to understand the influence of these strategies on physicians' symmetrical prescription behavior. Though, to the best knowledge of the authors, limited attention has been given to this topic, especially, the effect of marketing on physicians' symmetrical prescription behavior. Moreover, limited work has been carried out on customer relationships with pharmaceutical companies and physicians' personal traits with preference to marketing communication elements, and the impact of a corporate image with reference to the physicians' prescription decisions. This study will attempt to address this deficiency. The present study may help to provide a better understanding of marketing and medical tools, and their influence on physicians' symmetrical prescription behavior. Besides marketing and medical factors, we incorporate two important moderating variables, and examine the influence of corporate image and customer relationship in an association of medical and marketing factors, and the symmetry of physicians' prescription behavior in this study. The results of this study could be more conclusive to the pharmaceutical marketing and medical managers to improve their strategies and to make the best use of their financial and human resources for optimal gain of prescription business in an efficient and effective manner.

3. Academics Relevance

This research study has a unique academic perspective, which is relevant in offering comprehensive insight regarding the influential factors for the symmetry of physicians' prescription behavior within the context of Pakistani Pharmaceutical industry. The undertaken study examines the existing and novel factors that may influence the physicians' symmetrical prescription behavior. This research will examine a comprehensive analysis of exogenous, moderating factors on the physicians' symmetrical prescription behavior. In this research, besides marketing factors, we have also included medical factors that could be important for reshaping the symmetry of physicians' behavior. Finally, the undertaken study is unique in nature because it examines the effects of moderating variables such as corporate image and customer relationships in an association of medical & marketing factors, and the symmetry in physicians' prescription behavior [3,14,22]. Thus, the undertaken study could be a significant addition to the current literature regarding pharmaceutical marketing and physicians' behavior. Moreover, this research may also set a new horizon for future research studies, and provide the foundation for forthcoming literature regarding antecedents of physicians' symmetrical prescription behavior.

4. Previous Literature and Conceptual Framework

The Physicians' Prescription Behavior Definition

The symmetry in physicians' prescription behavior is multi-faceted and has a broad concept that includes several dimensions; however, in this study we focus on the symmetry of physicians' prescription behavior as an "adoption" [23,24]. The American Marketing Association has defined "adoption" as: "new product awareness, gathering information, developing positive attitudes towards the product, testing it in some direct or indirect way, finding satisfaction in the trial and adopting the product into a standing usage or repurchase pattern [3,18,25]". The process of diffusion is also signified as the process of adoption; it is a process in which novel ideas and products become norms and are accepted by society. According to Rogers [26], and Masood et al. [8], adoption is a social process, and social tendencies originate from the process of adoption. In pharmaceutical marketing, it is a process of influential activities to change the behavior, attitude, knowledge, and symmetry patterns of prescriptions of physicians [1,27,28].

5. Pharmaceutical Medical and Marketing Tools

Medical Literature & Journal Advertising

The medical literature has played a significant role in shaping the symmetry of the prescription behavior of physicians for a specific drug. Research articles published in credible journals such as

British medical journal, Lancet, JAMA, and American journal of medicine etc., set or change the direction of physicians' symmetrical prescriptions [27]. Pharmaceutical companies spend millions of dollars to conduct different large-scale studies, and favorable results of a drug to publish in credible journals to influence the symmetry of physicians' prescription behavior [25,29]. According to Stafford et al. [30], and Hersh et al. [31], the pharmaceutical companies fund multicenter trials, and significant medical breakthroughs are being brought to the attention of the general public and the medical community to change the perception of a given disease area and its cure. Interestingly, nobody knows who is behind these trials, and pharmaceutical companies usually fund the medical universities and research centers to obtain favorable results. Then, these results are utilized to influence the symmetry of physicians' prescription behavior [1,15,32]. Medical Journals' advertising is another significant window of opportunity to enhance the awareness of a drug. According to Loden and Liebman [33], the pharmaceutical industry spent $278.9 million on medical journal advertising in 1999, and that figure reached over $752 million in 2002. According to Leffler [34], journal advertising plays a double role: it enhances awareness, and also increases the influence on physicians; thus, medical journal advertising has been revealed as pro-competitive and lessening drug price following entry of a new drug. Rizzo [35] and Liebman [36] have studied the effectiveness of journal advertising, and concluded that journal advertising disseminates the right message and implementation that establish drug acceptance more than advertising expenditure. Several research studies have demonstrated the superiority and effectiveness of medical journals in terms of influence on the symmetry of physicians' prescription behavior [37,38]. Hence, the following hypothesis is formulated on the basis of previous research studies:

H1. *Medical literature & Journals advertising has a significant positive influence on physicians' prescription behavior.*

6. Scientific Activities

Pharmaceutical companies sponsor symposia, lectures, local speaker programs, doctors group meetings, ward presentations, and roundtable discussions to individual doctors and hospitals to influence the symmetry of physicians' prescription behavior [3,16]. There is sufficient evidence that these scientific activities are part of the medical and marketing activities of pharmaceutical companies, which definitely pays off for the pharmaceutical industry in terms of strong relationships and prescription business from the doctors and hospitals [10,15,39]. According to Rahman et al. [25], Ziegler [40], and Chren and Landefeld [41], doctors who are sponsored to attend these scientific activities are more convincing at writing the prescription of a drug of a sponsor company without doing a cost of benefit analysis of the drug. Pharmaceutical sponsored scientific activities include national and international conferences, training programs, and CME programs to the hospitals to include their drugs in the formulary [32,42,43]. Pharmaceutical companies sponsor individual physicians for international conferences; the objective of these sponsorships are multi-dimensional, such as the sponsored physicians have to get the recent development in the disease area, and then sponsored physicians are expected to deliver series of lectures to the junior doctors after returning to their home countries. However, the apparent objective is very positive and convincing, but pharmaceutical companies get additional prescriptions from the sponsored physicians, and on the other hand, the sponsored physicians play the role of advocator sand brand ambassadors of different drugs of sponsoring companies [10,44]. According to Bowman and Pearle [45], Lexchin [46], Grundy et al. [47], local and international CMEs play a vital role in influencing the symmetry of physicians' prescription behavior. Studies have revealed that after availing sponsorships, the physicians increase prescription rates by up to three times. Scheffer [24], and Lieb and Koch [48] pointed out those physicians who did not avail the companies' sponsorships tend to fairly prescribe the medicines on merit as compared to sponsored doctors. According to Vicciardo [49], Wazana [27], and Holmer [50], the CMEs activities are more linked with the pharmaceutical marketing objectives rather than medical activities, and companies usually cross ethical barriers in the name of CMEs. Hence, a substantial amount of literature has demonstrated that these scientific activities are more linked to the marketing of drugs, and CMEs have a significant influence on the symmetry of physicians' prescription behavior [1,23,51–53]. Thus, the following hypothesis is formulated on the basis of previous research studies:

H2. *Scientific activities have a significant positive influence on the symmetry of physicians' prescription behavior.*

7. Medical Representatives' Effectiveness

Pharmaceutical marketing is a kind of personal selling in which Medical representatives' detailing has a significant influence on the symmetry of physicians' prescription behavior [37,54]. Medical detailing is the most influential factor in pharmaceutical marketing because it is a continuous human interaction with the physicians at regular intervals of time. Moreover, Medical representatives' competence and skills also play a significant role in changing the physicians' symmetrical prescription behaviors in an effective and efficient manner [23,55]. Physicians tend to be more eager to see MRs because of new information about drugs, new developments in a given disease area, and to get free samples and gifts; thus, substantial literature has demonstrated the effectiveness of Medical representatives and influence on the physicians' symmetrical prescription behavior [3,27,56]. The effective sales call has three components, i.e., solid message content, well-utilized resources, and clear message delivery. The components of the sales message are an indication of a drug, side effects, dosage, efficacy, and competitive data [1,56]. Important research was carried out by the Wazana [27] to evaluate the cost and benefit of Physician-MR interaction. He concluded that "interactions with pharmaceutical representatives were found to impact the prescribing practice of residents and physicians in terms of prescribing cost, non-rational prescribing, awareness, preference and rapid prescribing of new drugs, and decreased prescribing of generic drugs" [27]. Caudill et al. [57], and Narayanan et al. [20] concluded that medical representatives' influence on physician is directly correlated to the level of credibility and effectiveness of MRs. Several research studies have demonstrated that medical representative's positive attributes such as knowledge about drugs, interpersonal skills, selling skills, and relationship with the physicians have a significant influence on the doctors' symmetrical prescription behavior [58,59]. Hence, a substantial amount of literature has confirmed the influential impact of MRs on the symmetry of physicians' prescription behavior [1,6,60]. Thus, the following hypothesis is formulated on the basis of previous research studies:

H3. *Medical representatives' effectiveness has a significant positive influence on the symmetry of physicians' prescription behavior.*

8. Promotional Material (Samples & Gifts)

Several researchers have carried out studies on the influence of samples and promotional material on the symmetry of physicians' prescription behavior. They have revealed in their findings that samples actually provide access to physicians' chambers, and give confidence to the medical representatives as well [1,58,59,61]. According to Findlay [62], pharmaceutical companies spend a huge portion of their marketing cost on free samples; he estimated companies have spent $7.2 billion in a single year. According to Chew et al. [63], physicians can deviate from their choice of drugs while they give these free samples to their patients. Thus, it is indicated that the free samples have a definite influence on physicians' symmetrical prescription behavior. Similar results have been reported in another study stating that family physicians give these free samples to their patients to help reduce the cost of filling a prescription [64–66]. The free samples are given for several reasons: to compete with other drugs, to launch a new product, to establish an efficacy of a new drug, to introduce a drug to a new doctor, to change the image of a drug, and to enhance the familiarity and demand of drug. Extensive literature has demonstrated that samples have a significant impact, which influences the symmetry of prescription behavior of a physician [37,67]. Apart from free samples, pharmaceutical companies also use inexpensive gifts as reminders, including prescription pads, pens, paperweights, calendars, mountings, table organizers, wall clocks, diaries, and other gifts [27,68]. These gifts play the vital role of giving a constant reminder of a drug beyond the sales calls to the physician; thus, these reminder gifts play a vital role to enhance the physicians' symmetrical prescription for a certain drug [23,24,69]. Besides these inexpensive gifts, pharmaceutical companies also give some valuable gifts to the physicians for the patients' welfare such as stethoscopes, weighing machines, BP apparatuses, and Blood sugar test machines with strips, water dispensers, drug refrigerators, and other valuable gifts to buy a physicians' time for detailing and for maintaining good relationships. Research studies have demonstrated that all kinds of gifts give opportunities to pharmaceutical companies to influence the symmetry of physicians' prescription behavior [1,51,70]. Thus, the following hypothesis is formulated on the basis of previous research studies:

H4. *Promotional material has a significant positive influence on the symmetry of physicians' prescription behavior.*

9. Personal Obligations

Pharmaceutical companies usually pay doctors' travel expenses directly, and sometimes pay even the doctor's family's travel expenses [55]. The gifts presented to doctors range from stationery and office-related gifts with minimal value to more personal and innovative gifts such as household related gifts, overseas trips, and air-conditioners; but the huge monetary value gifts such as travel tickets and vacations are less common than inexpensive ones such as pens, notepads, and coffee mugs in pharmaceutical promotion [37,71]. Furthermore, one of the most common pharmaceutical promotional gifts is material for patient care and gifts unrelated to medicine practice [14,72]. There are several and common promotional practices intended by pharmaceutical companies to promote their products to doctors, such as paying for vacations or the travel expenses of doctors, offering them valuable gifts, lavish meals and entertainment, giving them cash commissions for prescribing a specific drug, money for drug trials, free medical samples, and promotional materials, as well as funding Continuing Medical Education (CME) and honoraria for teaching or speaking in such activities [16,27]. Physicians agree that these sorts of personal gifts are the indirect efforts of pharmaceutical companies to augment the symmetrical prescription business [1,73,74]. Personalized pharmaceutical marketing enhances the prescriptions as well as sponsorships for education and recreational activities and expensive gifts. Studies have demonstrated that physicians are more concerned about the ethical norms compared to medical residents (RMOs) [15,32]. However, physicians are more tilted towards local and foreign sponsorships, donations, and expensive give aways at the beginning of their careers that are provided by the pharmaceutical industry [23,24,75,76]. Hence, the following hypothesis is formulated on the basis of previous research studies:

H5. *Personal gifts have a significant positive influence on the symmetry of physicians' prescription behavior.*

10. The Effect of Direct-to-Consumer Advertising (DTCA)

Pharmaceutical companies are using the direct-to-consumer-advertising (DTCA) channel to provide the prescription information for their existing and new products to consumers (patients) [5,77,78]. The objective of pharmaceutical companies is to persuade patients to ask their physicians for the prescription of these drugs. Thus, pharmaceutical companies indirectly influence the symmetry of physicians' prescription behavior [6,79,80]. Previous literature adequately suggested that physicians have a tendency to accept patients' requests. According to Herzenstein et al. [81], and Gupta et al. [23], the DTCA has a positive impact on patients to ask for a specific drug from their physicians, and this may increase the chances of symmetrical prescription business. However, studies are also available in which physicians are not influenced because of the safety profile, inappropriateness of drug, or availability of other, less expensive drugs [82]. Besides the affirmative effects of DTCA, there is a negative impact which is also seen when patients assess the effects of adoption and diffusion through DTCA. Nair et al. [83], and Pirisi [84] have evaluated this effect when physicians are forced to prescribe any specific drug that is not appropriate to the patient; thus, in this manner, patients have to pay the price (side effects, expensive drug, overdose, misuse etc.) of DTCA influence [2,37,85,86]. According to Manchanda et al. [87], and Pirisi [84], the American Association of pharmaceutical scientists stated that 91% physicians felt pressure to fulfill patients' request. Thus, the following hypothesis is formulated on the basis of previous research studies:

H6. *Direct-to-consumer-advertising (DTCA) has a significant positive influence on the symmetry of physicians' prescription behavior.*

11. Moderating Influence: Corporate Image and Customer Relationship

Corporate Image

Most organizations are aware that a promising corporate image can distinguish a company with a credible reputation [88–92]. Thus, the organization cannot rely solely on their products and brands as a measure of added value and effective differentiation. This has happened partly due to the convergence of the standards of product quality and capabilities, and the increasing requirement of transparency and accountability. According to Ahmed et al. [1], Anderson and Sullivan [93], Ettenson and Knowles [94], and Dowling [95], building an affirmative corporate image is deemed by many organizations to be an augmented and effective means of differentiation that provides a unique competitive advantage from rival organizations. The building of an appropriate corporate image through individuality entails a significant investment in terms of management efforts, time, and financial capitals; thus, it requires quantifiable and clear returns on investment. The existing literature has demonstrated that some quantifiable returns could be created through a corporate image; the corporate image may help to increase revenue and auxiliary support to the new product development [96–98]. According to Newell and Goldsmith [99], corporate image may also enhance financial relations; Smith [100] has argued that corporate image may improve employees' recruitment, retention, and relations. According to Dowling [101], corporate image is helpful for faster recovery in crises periods; Gotsi and Wilson [102] have pointed out that the corporate image is a great source for the development of emotional values that may improve the brand value as well. Intense transformations in the industry have propelled increased organization interest in the opinions held by the key stakeholders and the favorable corporate image value. Several pharmaceutical organizations now also agree on the prominence of recognizing the key features that are used by numerous stakeholders to craft a good corporate image [58,59,103]. According to Flavian et al. [104], the success of a promising corporate image is based on effective marketing communication strategies that attract existing and new customers. Physicians are more likely to prescribe medicines that are manufactured by to a prestigious company; thus, physicians are more comfortable prescribing branded drugs [105,106]. In the case of Pakistani pharmaceutical industry, doctors are easily tilted towards branded or copy brands that belong to a reputable national and multinational company [3,25]. Hence, the following hypotheses are formulated on the basis of previous research studies:

H7A. *Corporate image has a significant moderating effect between medical literature & Journal advertising, and physicians' symmetrical prescription behavior.*

H7B. *Corporate image has a significant moderating effect between scientific activities and, physicians' symmetrical prescription behavior.*

H7C. *Corporate image has a significant moderating effect between Medical representatives' effectiveness, and physicians' symmetrical prescription behavior.*

H7D. *Corporate image has a significant moderating effect between promotional material, and physicians' symmetrical prescription behavior.*

H7E. *Corporate image has a significant moderating effect between personal obligations, and physicians' symmetrical prescription behavior.*

H7F. *Corporate image has a significant moderating effect between direct-to-consumer-advertising, and physicians' symmetrical prescription behavior.*

12. Customer Relationship

Customer relationship plays a vital role between the pharmaceutical company and the physician, and its importance has been increasing for years. The marketing and promotion of a drug starts and ends at the level of physician. With pharmaceutical products, the consumer (patient) is not a direct customer of an organization; rather, the physician is the core customer for the pharmaceutical industry [14,22,107]. It is very interesting that, on one hand, the patient is not the direct consumer; however, on the other hand, the actual customer (physician) has to believe and rely on the information that is provided by the pharmaceutical company. Thus, it has a unique relationship that cannot be observed in any other industry [43,108–110]. Therefore, in pharmaceutical marketing, the customer relationship management (CRM) process is very significant, and marketing communication mix plays an important role in order to strengthen the relationship between physician and pharmaceutical company. Hence, all the national and transnational companies devise their marketing communication campaigns to enhance and sustain good relationships with physicians across the globe [6,60]. The long-run relationship not only strengthens business relationships, but it is also important to understand the physicians' psyche and behavioral intentions towards symmetry in the prescription business [64,111]. The customer relationship marketing has now become paramount in every industry, and in the pharmaceutical industry, it has a distinctive importance: relationship marketing is an on-going process to engage physicians in the corporate activities and programs to enhance the mutual economic benefits at minimal cost [1,9,27]. Thus, in the pharmaceutical industry, companies have devised effective relationship-building campaigns with physicians that are helpful in retaining their competitive advantage among today's intense competition. Medical detailing, company-sponsored conferences, CMEs, drug trials, involvement in scientific activities, personalized gifts, etc. are major and effective tools for developing customer relationships with physicians. Pharmaceutical companies are enjoying a long-lasting relationship with physicians in the form of the augmented and symmetrical prescription business that is based on credibility and trust [3,15]. Thus, the following hypotheses are formulated on the basis of previous research studies:

H8A. *Customer relationship has a significant moderating effect between medical literature & Journal advertising, and symmetry of physicians' prescription behavior.*

H8B. *Customer relationship has a significant moderating effect between scientific activities, and symmetry of physicians' prescription behavior.*

H8C. *Customer relationship has a significant moderating effect between Medical representatives' effectiveness, and symmetry of physicians' prescription behavior.*

H8D. *Customer relationship has a significant moderating effect between promotional material, and symmetry of physicians' prescription behavior.*

H8E. *Customer relationship has a significant moderating effect between personal obligations, and symmetry of physicians' prescription behavior.*

H8F. *Customer relationship has a significant moderating effect between direct-to-consumer-advertising, and symmetry of physicians' prescription behavior.*

Based on the above discussions and the literature, we have developed and articulated the following conceptual model of our research study (Figure 1):

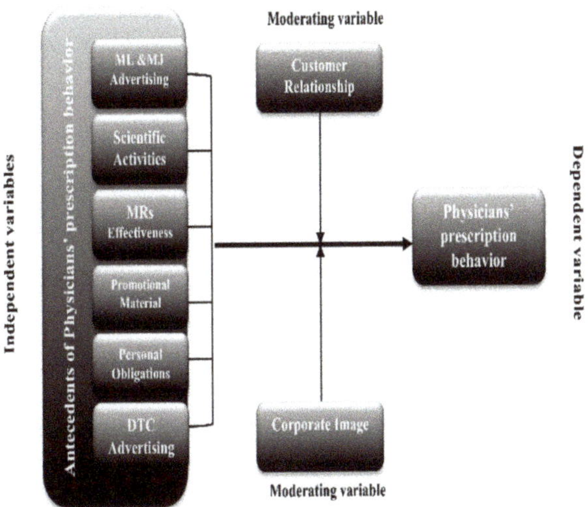

Figure 1. Conceptual Model of the Study based on previous literature.

13. Material and Methods

13.1. Estimation Techniques

We have employed several estimation techniques to analyze the results, such as descriptive statistics to analyze the characteristics of central tendencies and measure of location of variables, reliabilities (factor loading for individual items, and Cronbach's alpha, composite reliability & average variance extracted for factors) and validities (convergent and discriminant validities to confirm the construct validity) analyses to ascertain items and constructs, and SEM-based multivariate approaches, including exploratory factor analysis to validate and retain the items and variables such as rotated component matrix, KMO and Bartlett's Sphericity tests, total variance explained, and anti-image matrix. We used confirmatory analysis to ascertain the theoretical constructs and factors. Thus, we employed indices statistics to validate the hypothesized measurement and structural models. Finally, we have employed Hayes conditional process modeling to substantiate the direct and indirect influence of exogenous and moderating variables, such as moderating analysis, R-square increase, conditional indirect effect analysis, and 3D plotting, to validate the moderating influence. For this purpose, we used MS-Excel 10, SPSS 22, AMOS 22, and Hayes conditional process modeling software 2016.

13.2. Data Collection and Respondents' Profile

We sent out a total of 800 questionnaires to the general practitioners and medical specialists across Pakistan, and we received 740 responses with complete answers; 60 questionnaires were rejected due to incompleteness or inappropriate responses. Therefore, we achieved a 92.50% success rate, and according to MacCallum et al. [112], and Anderson and Gerbing [113], this was considered excellent. The outcomes shown in Table 1 demonstrated the respondents profiling, that shows that 428 (57.8%) responses were from male physicians, and 312 (42.2%) were received from female physicians. In terms of marital status, 295 (39.9%) respondents were single, 426 (57.6%) were married, and 19 (2.6%) respondents were divorced. As far as the age bracket of our respondents is concerned, 266 (35.9%) respondents were within the age bracket of 20–30 years, 164 (22.2%) within the age bracket of 30–40 years, 96 (13.0%) within the age bracket of 40–50 years, 124 (16.8%) within the age bracket of 50–60 years, and 90 (12.2%) were more than 60 years of age. Similarly, in terms of education, 326 (44.1%) received their medical graduate degrees, 227 (30.7%) received local post-graduate degrees, and 123 (16.6%) received foreign post-graduation, whereas, 64 (8.6%) received higher degrees such as a

Ph.D. in medicine. As far as the experience of the respondents is concerned, 201 (27.2%) had 1–5 years working experience, 218 (29.5%) had 5–10 years experience, 99 (13.4%) had 10–15 years working experience, 103 (13.9%) had 15–20 years of experience, and the remaining 109 (16.1%) respondents had more than 20 years working experience. Finally, in terms of the income of respondents, 121 (16.4%) physicians had an income ranging between 30–60 thousand Pakistani Rupee per month, 332 (44.9%) of 60–90 thousand, 154 (20.8%) between 90–120 thousand, 80 (10.8%) physicians earned between 120–150 thousand, and 53 (7.2%) respondents earned more than 150 thousand per month.

Table 1. Profile of Respondents.

Demographics		Frequency	Percent
	Male	428	57.8%
	Female	312	42.2%
	Single	295	39.9%
Marital Status	Married	426	57.6%
	Divorced	19	2.6%
	20–30	266	35.9%
	30–40	164	22.2%
Age (In Years)	40–50	96	13.0%
	50–60	124	16.8%
	More than 60	90	12.2%
	Graduation	326	44.1%
Education	Post-Graduation (Local)	227	30.7%
	Post Graduation (Foreign)	123	16.6%
	PhD degree	64	8.6%
	1–5	201	27.2%
	5–10	218	29.5%
Experience (In Years)	10–15	99	13.4%
	15–20	103	13.9%
	More than 20	119	16.1%
	30–60	121	16.4%
	60–90	332	44.9%
Income (In PKR 000)	90–120	154	20.8%
	120–150	80	10.8%
	More than 150	53	7.2%
Total–N		**740**	

Source: Authors' estimation.

14. Data Analysis and Results Estimations

In this section, we report the generated results and their interpretations from different software such as MS-Excel 10, SPSS 22, AMOS 22, and Hayes conditional process analysis. The following is a description of the results, which we extracted from the aforementioned software.

15. Descriptive Statistics of Initial Constructs

According to Ahmed et al. [114], and Huang et al. [115], the normality of data is a pre-requisite for the analysis of SEM-based approaches. Therefore, we have transformed our considered data sample into z-scores and used descriptive statistics. The results of Table 2 demonstrate that all the construct values of Kurtosis ranged between (−3 to +3), and the values of standard deviations and Skewness for all the items ranged between −1.5 to +1.5, which also validated that the data sample follows a normality pattern [3,116,117].

Table 2. Descriptive Statistics.

Statistics		PPB	MLJ	SAC	MRE	PMT	POB	DTCA	CIM	CRP
N	Valid	740	740	740	740	740	740	740	740	740
	Missing	0	0	0	0	0	0	0	0	0
Mean		3.8027	3.9757	3.9311	3.9662	3.9324	3.8797	4.0270	3.9311	4.0608
Std. Deviation		1.0832	1.1227	1.1008	1.1182	1.1014	1.0726	0.9517	1.0137	1.0249
Skewness		−0.928	−1.010	−0.986	−1.004	−0.987	−0.973	−1.083	−0.940	−1.105
Std. Error of Skewness		0.090	0.090	0.090	0.090	0.090	0.090	0.090	0.090	0.090
Kurtosis		0.359	0.376	0.450	0.389	0.447	0.575	1.472	0.664	0.923
Std. Error of Kurtosis		0.179	0.179	0.179	0.179	0.179	0.179	0.179	0.179	0.179

PPB = Physicians' prescription behavior; MLJ = Medical literature & Journal advertising; SAC = Scientific activities; MRE = Medical representatives' effectiveness; PMT = Promotional material; POB = Personal obligations; DTC = Direct-to-consumer advertising; CIM = Corporate image; CRP = Customer relationship. Source: Authors' estimation.

16. Reliabilities and AVE Analysis

The outcomes in Table 3 demonstrate that the values of Cronbach's and composite reliabilities range between 0.80–0.90, which meets the minimum threshold criteria (>0.6) [114,118]. The outcomes of Table 3 further demonstrate that the values of factor loadings range between 0.70–0.95, which also meets the criterion of discriminant validities [114,116]. Similarly, the values of average variance extracted (AVE) are greater than 0.50 in all items; hence, the condition of convergent validity has also been fulfilled, which is also a pre-requisite for the SEM, EFA, and CFA approaches [119].

Table 3. Reliabilities and Average Variance Extracted.

Factors	Items	FL	CA	CR	AVE
Physicians' prescription behaviour	PPB1 PPB2 PPB3	0.804 0.897 0.81	0.837	0.876	0.702
Medical literature & Journal advertising	MLJ1 MLJ2 MLJ3 MLJ4	0.924 0.88 0.959 0.718	0.870	0.928	0.766
Scientific activities	SAC1 SAC2 SAC3 SAC4	0.807 0.92 0.822 0.986	0.884	0.936	0.786
Medical representative effectiveness	MRE1 MRE2 MRE3	0.854 0.835 0.838	0.842	0.880	0.709
Promotional material	PMT1 PMT2 PMT3	0.780 0.884 0.807	0.824	0.864	0.680
Personal obligations	POB1 POB2 POB3	0.831 0.872 0.776	0.826	0.867	0.684
Direct-to-consumer-advertising (DTCA)	DTC1 DTC2 DTC3	0.845 0.874 0.795	0.838	0.877	0.703
Corporate Image	CIM1 CIM2 CIM3	0.766 0.841 0.921	0.843	0.882	0.714
Customer Relationship	CRP1 CRP2 CRP3	0.768 0.819 0.894	0.827	0.868	0.687

PPB = Physicians' prescription behavior; MLJ = Medical literature & Journal advertising; SAC = Scientific activities; MRE = Medical representatives' effectiveness; PMT = Promotional material; POB = Personal obligations; DTC = Direct-to-consumer advertising; CIM = Corporate image; CRP = Customer relationship; Source: Authors' estimation.

17. Exploratory Factor Analysis—EFA

For the reduction of the sample data, we employed exploratory factor analysis; the loaded sample data is further divided into factors and constructs. Thus, the EFA recruited the comparable item and factors. The EFA has the tendency and efficiency to shrink large amounts of data into meaningful, small segments. According to Emory and Cooper [120], the EFA approach may support the researcher in examining the precision of variables. The results in Table 4 demonstrate the outcomes of the rotated component matrix that validated the factors and items of the considered data sample. This research study comprises nine factors and twenty-nine items of these factors, for which PPB has three items, and MJA and scientific activities (SAC) have three constructs each. Similarly, MRE, PMT, POB, direct-to-consumer advertising (DTC), corporate image (CIM), and customer relationship have three items each factor. The outcomes of Table 4 further demonstrated that the value of factor loading for each construct is greater than 0.50, which establishes that our considered factors and items are valid [3,118,121].

Table 4. Rotated Component Matrix [a].

Factors	Items	Factor Loadings of Components								
		PPB	MLJ	SAC	MRE	PMT	POB	DTC	CIM	CRP
Physicians' prescription behavior	PPB1	0.804								
	PPB2	0.897								
	PPB3	0.81								
Medical literature & Journal advertising	MLJ1		0.924							
	MLJ2		0.88							
	MLJ3		0.959							
	MLJ4		0.718							
Scientific activities	SAC1			0.807						
	SAC2			0.92						
	SAC3			0.822						
	SAC4			0.986						
Medical representatives' effectiveness	MRE1				0.854					
	MRE2				0.835					
	MRE3				0.838					
Promotional material	PMT1					0.780				
	PMT2					0.884				
	PMT3					0.807				
Personal obligations	POB1						0.831			
	POB2						0.872			
	POB3						0.776			
Direct-to-consumer advertising (DTCA)	DTC1							0.845		
	DTC2							0.874		
	DTC3							0.795		
Corporate Image	CIM1								0.766	
	CIM2								0.841	
	CIM3								0.921	
Customer Relationship	CRP1									0.768
	CRP2									0.819
	CRP3									0.894

Extraction Method: Principal Component Analysis. Rotation Method: Varimax with Kaiser Normalization. [a] Rotation converged in 6 iterations. PPB = Physicians' prescription behavior; MLJ = Medical literature & Journal advertising; SAC = Scientific activities; MRE = Medical representatives' effectiveness; PMT = Promotional material; POB = Personal obligations; DTC = Direct-to-consumer advertising; CIM = Corporate image; CRP = Customer relationship. Source: Authors' estimation.

18. Kaiser Meyer Olkin (KMO) and Bartlett's Techniques

The Bartlett's sphericity test and Kaiser Meyer Olkin (KMO) analysis establish the sampling adequacy and sample data fitness. The outcomes of Table 5 demonstrate that the value of KMO is 0.772, which is regarded as fairly good, because Kaiser [121] pointed out the values of KMO ranging between 0.70–0.79 are considered good. The outcomes of Table 5 further demonstrate that the value of Bartlett's sphericity shows that the corresponding probability is less than 0.000; however, the minimum required threshold value is less than 0.50. Thus, the value of $p < 0.05$ demonstrates the correlation between the constructs, which shows the correlation amongst the items is significant and adequate, at 5% significance level.

Table 5. KMO and Bartlett's Test.

Kaiser-Meyer-Olkin Measure of Sampling Adequacy.		0.772
Bartlett's Test of Sphericity	Approx. Chi-Square	24,266.468
	Df	703
	Sig.	0.000

Source: Authors' estimation.

19. Total Variance Explained

We have considered nine factors, and cumulative percentages of variance determined the dispersion of variance of these factors. Moreover, the individual Eigenvalues should also be greater than one. The outcomes of Table 6 establish that the cumulative variance of nine factors is 74.08%, which is well above the minimum required threshold value of 50%. Table 6 further exhibits that the total Eigenvalues of each factor is more than one, which also confirms the amount of variance among the potential factors. The results suggest retaining all the nine variables, and now we can proceed to further analysis.

Table 6. Total Variance Explained.

Factors	Initial Eigenvalues			Extraction Sums of Squared Loadings			Rotation Sums of Squared Loadings		
	Total	% of Variance	Cumulative %	Total	% of Variance	Cumulative %	Total	% of Variance	Cumulative %
1	6.748	17.759	17.759	6.748	17.759	17.759	6.577	17.309	17.309
2	3.726	9.806	27.565	3.726	9.806	27.565	2.638	6.942	24.251
3	2.965	7.804	35.368	2.965	7.804	35.368	2.636	6.938	31.189
4	2.708	7.127	42.495	2.708	7.127	42.495	2.629	6.919	38.108
5	2.624	6.906	49.401	2.624	6.906	49.401	2.625	6.909	45.017
6	2.564	6.749	56.150	2.564	6.749	56.150	2.611	6.872	51.889
7	2.425	6.383	62.533	2.425	6.383	62.533	2.601	6.845	58.734
8	2.293	6.034	68.567	2.293	6.034	68.567	2.570	6.762	65.496
9	2.093	5.508	74.075	2.093	5.508	74.075	2.517	6.625	72.120

Extraction Method: Principal Component Analysis. Source: Authors' estimation.

20. The Anti-Image Matrix

For the selection and de-selection of variables, we employed an anti-image correlation matrix (AICM) approach. If any variable exhibits measures of sampling adequacy (MSA) greater than 0.5 in the diagonal of the anti-image correlation matrix (AICM), then that variable will be retained. The results of Table 7 show that all factors have a MSA value of more than 0.5 in the diagonal of AICM; therefore, we will retain all variables.

Table 7. Anti-image Matrices.

	Factors	PPB	MLJ	SAC	MRE	PMT	POB	DTCA	CIM	CRP
Anti-image Correlation	PPB	0.909 [a]	−0.093	−0.245	−0.078	−0.561	−0.056	0.084	−0.015	−0.044
	MLJ	−0.093	0.946 [a]	0.039	−0.276	−0.070	−0.272	0.078	−0.182	0.034
	SAC	−0.245	0.039	0.877 [a]	−0.684	−0.109	−0.135	−0.055	−0.043	0.049
	MLJ effectiveness	−0.078	−0.276	−0.684	0.870 [a]	0.095	−0.058	−0.024	0.128	−0.080
	PMT	−0.561	−0.070	−0.109	0.095	0.890 [a]	−0.356	−0.097	0.056	0.037
	POB obligations	−0.056	−0.272	−0.135	−0.058	−0.356	0.944 [a]	0.036	−0.029	0.021
	DTCA	0.084	0.078	−0.055	−0.024	−0.097	0.036	0.706 [a]	−0.394	−0.662
	CIM	−0.015	−0.182	−0.043	0.128	0.056	−0.029	−0.394	0.826 [a]	−0.259
	CRP	−0.044	0.034	0.049	−0.080	0.037	0.021	−0.662	−0.259	0.732 [a]

[a] Measures of Sampling Adequacy (MSA). PPB = Physicians' prescription behavior; MLJ = Medical literature & Journal advertising; SAC = Scientific activities; MRE = Medical representatives' effectiveness; PMT = Promotional material; POB = Personal obligations; DTC = Direct-to-consumer advertising; CIM = Corporate image; CRP = Customer relationship. Source: Authors' estimation.

21. Confirmatory Factor Analysis—CFA

The confirmatory factor analysis is a direct and suitable approach for checking the measurement model in SEM-based approaches. The CFA confirms the prior theories and also determines whether our considered data sample is fit for the hypothesized measurement model [3,118]. In our measurement model, we have validated the factors for influencing the symmetry of PPB such as medical literature & journal advertising (MLJ), SAC, MRE, PMT, Personal obligations (POB), and DTC. Furthermore, we considered CIM and customer relationship (CRP) as moderating variables between exogenous and endogenous variables. In our measurement model, we considered twenty-nine items among nine factors, and fix the sample data between observed and unobserved variables [3,117]. The results of Table 3 show that the factor loadings of all the items range between 0.70–0.95, establishing that our overall measurement model is acceptable; the factor loading values also confirmed the discriminant validities of the hypothesized measurement model [3]. The outcomes of Table 8 show that the values of fit-indices (RNI:0.96, CFI:0.97, IFI:0.97, GFI:0.96, TLI:0.96, NFI:0.91, PNFI:0.86, PCFI:0.84 & RMSEA:0.003) are within the required threshold criterion for the considered measured model. Hence, it was concluded that our hypothesized measurement model is acceptable [114,118].

Table 8. Models Fit Statistics.

Goodness of Fit Measures	Absolute Fit Indices			Relative Fit Indices			Non-Centrality-Based Indices			Parsimonious Fit Indices	
	χ^2/df	Probability	GFI	NFI	IFI	TLI	CFI	RMSEA	RNI	PCFI	PNFI
Measurement Model	3.11	0.0043	0.96	0.91	0.97	0.96	0.97	0.003	0.96	0.84	0.86
Structural Model	3.25	0.0048	0.97	0.93	0.99	0.98	0.99	0.004	0.97	0.87	0.88
Criterion (Threshold values)	<5.0	<0.05	>0.95	>0.90	>0.95	>0.95	>0.95	<0.05	>0.95	>0.75	>0.75

Note: TLI = Tucker-Lewis Index; χ^2/d = Relative Chi square; GFI = Goodness of Fit Index; RMSEA = Root mean squared error of approximation; CFI = Comparative fit index; NFI = Normed fixed index; IFI = Incremental fixed index; RNI = Relative Non-centrality Index; PNFI = Parsimony-adjusted normed fit index; PCFI = Parsimonious-adjusted fit index. Source: Authors' estimation.

22. Structural Equation Modeling—SEM

For the estimations of factors for the (PPB), we employed the SEM-based structural model. We incorporated six factors pertaining to marketing and medical dimensions, namely, medical literature & journal advertising (MLJ), SAC, medical representatives' effectiveness (MRE), PMT, personal gifts (POB), and DTC. Furthermore, we considered CIM and customer relationship (CRP) as moderating variables between exogenous and endogenous variables. In our measurement model, we considered twenty-nine items from among nine factors, and fixed the sample data between observed and unobserved variables [3,117]. The outcomes of Table 8 demonstrated that based on our proposed research hypotheses, the outcomes of the structural model revealed that our whole model is accepted under the values of minimum threshold criterion of fit-indices. The results of Table 8 show that the values of fit-indices ((RNI:0.97, CFI:0.99, IFI:0.99, GFI:0.97, TLI:0.98, NFI:0.93, PNFI:0.88, PCFI:0.87 & RMSEA:0.004) are within the required threshold criterion for the considered structured model. Hence, it was concluded that our structural model is accepted as a suitable means to asses the symmetry in PPB [114,118].

23. Hypothesized Direct Relationship

For the interpretation of the direct relationship between marketing & medical factors and the symmetry of (PPB), we employed standardized regression weights. We ascertained the direct association of marketing & medical factors: medical literature & journal advertising (MLJ), SAC, MRE, PMT, personal gifts (POB), and DTC on the (PPB). The outcomes of Table 9 show that all the six null hypotheses (H1–H6) have been placed in the rejection region, as corresponding T-values are greater than 2, and p-values are less than 0.05. Thus, it was concluded that medical literature & journal advertising (MLJ), SAC, MRE, PMT, personal obligations (POB), and DTC have a significant positive influence on the symmetry of (PPB). The outcomes of Table 9 further reveal that PMT has the highest impact (0.915) on PPB, followed by the scientific activities (0.897) and personal obligations (0.888). However, MRE has the individual impact of (0.853), medical literature & journal advertising (MLJ) has (0.811), and DTC has an impact of (0.791) on the symmetry of PPB in the context of Pakistani healthcare industry.

Table 9. Hypothesized Direct Relationship.

Hypothesis	Variables	Regression Paths	Standardized Regression Weights (β)	SE	T	P	Decision
H1	Med Lit & Journal advertising	MLJ † → PPB	0.811	0.015	54.09	0.000	**Supported**
H2	Scientific activities	SAC † → PPB	0.897	0.012	74.55	0.000	**Supported**
H3	Medical Rep effectiveness	MRE † → PPB	0.853	0.013	64.37	0.000	**Supported**
H4	Promotional material	PMT † → PPB	0.915	0.011	81.46	0.000	**Supported**
H5	Personal obligations	POB † → PPB	0.888	0.013	63.97	0.000	**Supported**
H6	Direct-to-consumer advertising	DTC † → PPB	0.791	0.014	44.11	0.000	**Supported**

Note: † = Predictor; DV = PPB = Physicians' prescription behavior; MLJ = Medical literature & Journal advertising; SAC = Scientific activities; MRE = Medical representatives' effectiveness; PMT = Promotional material; POB = Personal obligations; DTC = Direct-to-consumer advertising; CIM = Corporate image; CRP = Customer relationship. Source: Authors' estimation.

24. Moderating Effect of CIM and CRP (Moderation Analysis)

The results of Table 10 show that moderating variables such as CIM and customer relationship (CRP) in association with the exogenous variables for instance medical literature & journal advertising (MLJ), SAC, MRE, PMT, personal gifts (POB), DTC, and symmetry of (PPB). The multiplicative effect of the MLJ and the moderating variable (CIM) exhibited a significant effect because the corresponding probability is less than 0.05. Hence, the null hypothesis H7A has been placed in rejection region, and the decision is in support of the statement. Similarly, the null hypotheses H7B and H8B were rejected, and it was concluded that the moderator CIM and CRP have a significant impact in the association of SAC, and the symmetry of PPB because the corresponding p-values are less than 0.05. The null hypothesis H8C has also been rejected, since the corresponding p-values are less than 0.05 ($p < 0.05$); hence, the moderator CRP has a significant effect on the relationship of MRE and PPB.

Similarly, the null hypotheses H7E and H7F have also been placed in the rejection region ($p < 0.05$); hence, the moderator CIM has a significant effect on the relationship between POB and PPB, and DTC and PPB. However, the remaining hypotheses, i.e., H8A, H7C, H7D, H8D, H8E, and H8F, do not lie in the rejection region because the corresponding probabilities are greater than $p > 0.05$. Hence, it was concluded that the moderating variable customer relationship (CRP) does not have a significant impact between MLJ and PPB, and similarly, the moderating variable CIM does not have a significant impact on MRE and PPB, the moderating variables such as CIM and CRP do not have significant impact on PMT and PPB, and POB and PPB, and CRP does not have significant impact on the relationship between DTC and PPB.

Table 10. Moderating Effect of CIM and Customer Relationship (CRP).

Hypotheses	Moderators	Moderation	Coefficient	SE	T	P *	LLCI	ULCI
		Moderating Effect of the CIM and CRP b/w MLJ and PPB						
H_{7A}:	CIM	MLJ × CIM	−0.0343	0.0122	−2.83	0.0049	−0.0582	−0.0105
H_{8A}:	CRP	MLJ × CRP	−0.0208	0.0120	−1.74	0.0831	−0.0443	0.0027
		Moderating Effect of the CIM and CRP b/w SAC and PPB						
H_{7B}	CIM	SAC ×CIM	−0.0263	0.0089	−2.94	0.0034	−0.0439	−0.0087
H_{8B}	CRP	SAC × CRP	−0.0343	0.0090	−3.80	0.0002	−0.0521	−0.0166
		Moderating Effect of the CIM and CRP b/w MRE and PPB						
H_{7C}:	CIM	MRE × CIM	−0.0201	0.0104	−1.93	0.0540	−0.0405	0.0003
H_{8C}:	CRP	MRE × CRP	−0.0349	0.0103	−3.40	0.0007	−0.0551	−0.0147
		Moderating Effect of the CIM and CRP b/w PMT and PPB						
H_{7D}:	CIM	PMT × CIM	−0.0131	0.0090	−1.45	0.1476	−0.0308	0.0046
H_{8D}:	CRP	PMT × CRP	−0.0143	0.0089	−1.60	0.1090	−0.0319	0.0032
		Moderating Effect of the CIM and CRP b/w POB and PB						
H_{7E}:	CIM	POB × CIM	−0.0224	0.0112	−1.99	0.0469	−0.0444	−0.0003
H_{8E}:	CRP	POB × CRP	−0.0136	0.0108	−1.25	0.2116	−0.0348	0.0077
		Moderating Effect of the CIM and CRP b/w DTC and PPB						
H_{7F}:	CIM	DTC × CIM	−0.0552	0.0157	−3.52	0.0005	−0.0859	−0.0244
H_{8F}:	CRP	DTC × CRP	−0.0029	0.0376	−0.76	0.9392	−0.0766	0.0709

Note: 'x' is known as the multiplicative sign; * denotes rejection of the hypotheses at 0.05 level ($p < 0.05$); DV = PPB = Physicians' prescription behavior; MLJ = Medical literature & Journal advertising; SA = Scientific activities; MRE = Medical representatives' effectiveness; PMT = Promotional material; POB = Personal obligations; DTC = Direct-to-consumer advertising; CIM = Corporate image; CRP = Customer relationship. Source: Authors' estimation.

25. R-Square Increment

According to Bolin [122], and Hayes [123], the increase in R-square occurs because of the catalyzing influence of the moderating variables. The outcomes of Table 11 show that an increase in R-squared occurs because of the interaction of moderating variables such as CIM and CRP in the association with the exogenous variables, for instance MLJ, SAC, MRE, PMT, POB, DTC, and PPB. The outcomes of moderation are substantiated from the higher values of F-statistics, and their corresponding probabilities are less than 0.05 ($p < 0.05$). Except for the moderation of CIM between MRE & PPB, PMT & PPB, and moderation of CRP between MLJ & PPB, PMT & PPB, POB & PPB, and DTC & PPB where $p > 0.05$.

Table 11. Increment In R-Square.

Moderation	R^2-Changed	F	df1	df2	P *
MLJ × CIM	0.0018	7.9810	1	736	0.0049
MLJ × CRP	0.0006	3.0113	1	736	0.0831
SAC × CIM	0.0010	8.6490	1	736	0.0034
SAC × CRP	0.0017	14.4272	1	736	0.0002
MRE × CIM	0.0006	3.7233	1	736	0.0540
MRE × CRP	0.0018	11.5312	1	736	0.0007
PMT × CIM	0.0002	2.1016	1	736	0.1476
PMT × CRP	0.0003	2.5755	1	736	0.1090
POB × CIM	0.0007	3.9637	1	736	0.0469
POB × CRP	0.0003	1.5635	1	736	0.2116
DTC × CIM	0.0042	12.4019	1	736	0.0005
DTC × CRP	0.0000	0.0058	1	736	0.9392

Note: 'x' is known as the multiplicative sign; * denotes rejection of the hypothesis at 0.05 level ($p < 0.05$). Source: Authors' estimation.

26. Conditional Effect of CIM and CRP (Moderators)

Results of the conditional effect of moderators such as CIM and CRP on independent variables such as MLJ, SAC, MRE, PMT, POB, DTC, and PPB as dependent variable are shown in Table 12. The extreme left column shows the various quantitative values of moderating variables such as CIM and CRP, which are mean of ±1 SD from the mean that matches to the 25th, 50th, and 75th, a percentile of the spreading of the CIM and CRP measures at 0.05 significance level. The results of conditional effect proposed that the independent variables such as MLJ, SAC, MRE, PMT, POB, DTC, and PPB at two measures of moderators (CIM & CRP) is significant, since the corresponding p-values are less than 0.05 [122,123], except for the conditional moderation of CRP between DTC & PPB ($p > 0.05$).

Table 12. Conditional Effects.

Conditional effect of MLJ on PPB on different values of the CIM						
CIM	Effect	SE	T	P *	LLCI	ULCI
2.9173	0.8398	0.0184	45.55	0.0000	0.8036	0.8760
3.9311	0.8050	0.0157	51.22	0.0000	0.7742	0.8359
4.9448	0.7702	0.0214	36.00	0.0000	0.7282	0.8122
Conditional effect of MLJ on PPB on different values of the CRP						
CRP	Effect	SE	T	P *	LLCI	ULCI
3.0359	0.8223	0.0171	48.21	0.0000	0.7888	0.8558
4.0608	0.8010	0.0155	51.69	0.0000	0.7706	0.8314
5.0000	0.7815	0.0214	36.52	0.0000	0.7395	0.8235
Conditional effect of SAC on PPB on different values of the CIM						
CIM	Effect	SE	T	P *	LLCI	ULCI
2.9173	0.9195	0.0133	69.07	0.0000	0.8934	0.9456
3.9311	0.8928	0.0126	70.72	0.0000	0.8681	0.9176
4.9448	0.8662	0.0175	49.53	0.0000	0.8319	0.9005
Conditional effect of SAC on PPB on different values of the CRP						
CRP	Effect	SE	T	P *	LLCI	ULCI
3.0359	0.9279	0.0131	70.66	0.0000	0.9021	0.9537
4.0608	0.8927	0.0125	71.31	0.0000	0.8681	0.9173
5.0000	0.8605	0.0171	50.28	0.0000	0.8269	0.8941

Table 12. Cont.

Conditional effect of MRE on PPB on different values of the CIM						
CIM	**Effect**	**SE**	**T**	**P ***	**LLCI**	**ULCI**
2.9173	0.8677	0.0152	57.21	0.0000	0.8379	0.8975
3.9311	0.8474	0.0138	61.23	0.0000	0.8202	0.8745
4.9448	0.8270	0.0194	42.67	0.0000	0.7890	0.8650

Conditional effect of MRE on PPB on different values of the CRP						
CRP	**Effect**	**SE**	**T**	**P ***	**LLCI**	**ULCI**
3.0359	0.8853	0.0153	57.81	0.0000	0.8552	0.9154
4.0608	0.8495	0.0137	62.22	0.0000	0.8227	0.8763
5.0000	0.8168	0.0184	44.49	0.0000	0.7807	0.8528

Conditional effect of PMT on PPB on different values of the CIM						
CIM	**Effect**	**SE**	**T**	**P ***	**LLCI**	**ULCI**
2.9173	0.9259	0.0127	72.65	0.0000	0.9009	0.9509
3.9311	0.9126	0.0117	78.18	0.0000	0.8897	0.9356
4.9448	0.8994	0.0167	53.95	0.0000	0.8666	0.9321

Conditional effect of PMT on PPB on different values of the CRP						
CRP	**Effect**	**SE**	**T**	**P ***	**LLCI**	**ULCI**
3.0359	0.9271	0.0126	73.53	0.0000	0.9023	0.9518
4.0608	0.9124	0.0116	78.50	0.0000	0.8896	0.9352
5.0000	0.8989	0.0161	55.70	0.0000	0.8672	0.9306

Conditional effect of POB on PPB on different values of the CIM						
CIM	**Effect**	**SE**	**T**	**P ***	**LLCI**	**ULCI**
2.9173	0.9065	0.0154	58.83	0.0000	0.8763	0.9368
3.9311	0.8839	0.0148	59.86	0.0000	0.8549	0.9129
4.9448	0.8612	0.0214	40.26	0.0000	0.8192	0.9032

Conditional effect of POB on PPB on different values of the CRP						
CRP	**Effect**	**SE**	**T**	**P ***	**LLCI**	**ULCI**
3.0359	0.8985	0.0150	60.05	0.0000	0.8692	0.9279
4.0608	0.8846	0.0147	60.33	0.0000	0.8558	0.9134
5.0000	0.8719	0.0206	42.36	0.0000	0.8315	0.9123

Conditional effect of DTC on PPB on different values of the CIM						
CIM	**Effect**	**SE**	**T**	**P ***	**LLCI**	**ULCI**
2.9173	0.8652	0.0248	34.83	0.0000	0.8164	0.9140
3.9311	0.8054	0.0196	41.14	0.0000	0.7669	0.8438
4.9448	0.7455	0.0270	27.65	0.0000	0.6926	0.7984

Conditional effect of DTC on PPB on different values of the CRP						
CRP	**Effect**	**SE**	**T**	**P ***	**LLCI**	**ULCI**
3.0359	0.1314	0.0975	1.35	0.1784	−0.0601	0.3228
4.0608	0.1284	0.0929	1.38	0.1674	−0.0540	0.3109
5.0000	0.1257	0.1022	1.23	0.2189	−0.0748	0.3263

* denotes rejection of the hypotheses at 0.05 level ($p < 0.05$); Quantitative values for moderators are mean and ± 1 SD from the mean; Level of confidence for all confidence intervals in output: 95.00. Source: Authors' estimation.

27. Visualization of Conditional Effect

The visualizing conditional effect can be seen in Figure 2, in which the moderating variables CIM, and customer relationship (CRP), in the relationship of MLJ, SAC, MRE, PMT, POB, DTC, and PPB. According to the Figure 2, the 3D plots exhibit that the independent variables such as MLJ, SAC, MRE, PMT, POB, DTC, and PPB experienced a significant influence of moderation, while we inducted two moderating variables such as CIM and customer relationship (CRP). The results of Figure 2 shows that the effect of prescription behavior is changing with the different values of moderating variables (CIM & CRP), with the constant values for independent variables (MLJ, SAC, MRE, PMT, POB & DTC) during this moderating process. Bolin [122], and Hayes [123] pointed out that graphical representation is an essential condition to show the moderation process because it clearly demonstrates the impact of moderation. Figure 2 uses red for the independent variables (MLJ, SAC, MRE, PMT, POB & DTC), orange for the incorporated moderators (CIM & CRP), and green for the effect of symmetry of PPB. Hence, it is, finally, concluded from the 3D graphs of Figure 2, the moderating variable CIM has a cogent influence amongst the association of MLJ, SAC, and DTC, and PPB in the context of the Pakistani pharmaceutical industry. However, the moderating variable customer relationship (CRP) has a significant influence amongst the association of SAC and MRE, and the symmetry of PPB in the context of the Pakistani pharmaceutical industry.

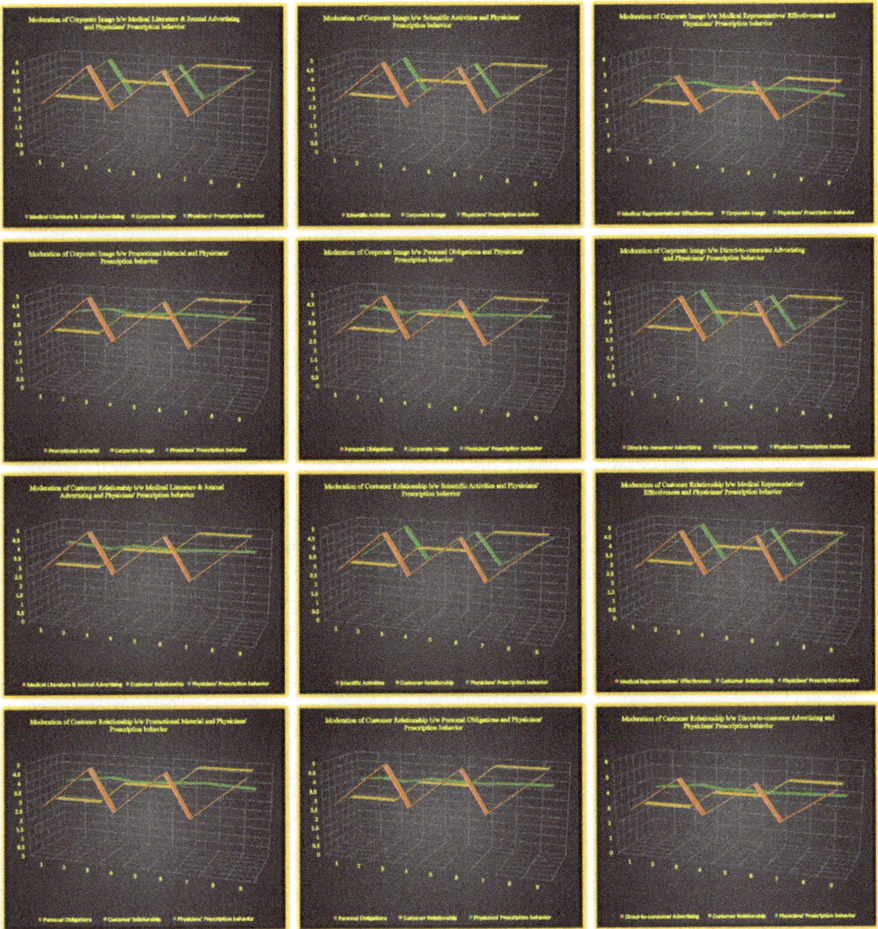

Figure 2. The visualizing conditional effect of moderators. Source: Authors' estimation.

28. Discussions

The objective of this study was to examine the factors which are important antecedents for the symmetry of Physicians' prescription behavior in the context of Pakistani pharmaceutical market.

Another objective was to ascertain the moderating variables such as corporate image and customer relationship between predictors and dependent variable i.e., the symmetry of physicians' prescription behavior. The results of the study demonstrated that independent variables such as medical literature & journal advertising have a positive and significant impact on Physicians' symmetrical prescription behavior, which is also substantiated the previous literature, for instance, Mulinari [15], Lotfi et al. [32], Rizzo [35], and DeJong et al. [37]. The outcomes of the study also showed that the scientific activities have a positive and significant influence on the symmetry of physicians' prescription behavior; this result also validated the previous literature [3,16,23,24,27,48]. The outcomes further demonstrated that the predictor, medical representatives' effectiveness, also has a strong and positive influence on the physicians' symmetrical prescription behavior, which is also in line with the previous research studies such as: De Ferrari et al. [6], Gupta et al. [23], Yeh et al. [55], Makowska [58], and Siddiqui et al. [59]. The results also demonstrated the positive and significant impact of promotional material on the symmetry of physicians' prescription behavior that is validated by the previous literature in works such as Ahmed, et al. [1], Yeh et al. [55], Makowska [58], Siddiqui et al. [59], Hurley et al. [64], and Katz et al. [70]. The outcomes of the study also concluded that the personal obligations have a direct and significant influence on physicians' prescription behavior; these results are in line with the previous research studies [14,16,27,37,55,71,124]. Finally, the results concluded that the direct-to-consumer advertising (DTCA) has a positive and cogent effect on the physicians' symmetrical prescription behavior; these results also certified the previous literature, such as Kesselheim et al. [2], Khan et al. [5], De Ferrari et al. [6], Alosaimi et al. [78], Osinga et al. [79], Shalowitz et al. [80], Gonül et al. [86]. The outcomes of moderating variables concluded that the corporate image has a significant influence between predictors such as Medical literature & Journal advertising, scientific activities, and direct-to-consumer advertising (DTCA), and endogenous variable i.e., physicians' symmetrical prescription behavior. These results are consistent with previous research studies such as Lieb and Scheurich [88], Brown et al. [96], Balmer and Greyser [105], Jaffe and Nebenzahl [106]. The outcomes of the moderator, customer relationship, showed evidence that the customer relationship has a significant impact amongst the association of scientific activities and Medical representative effectiveness, and the symmetry of PPB in the context of the Pakistani pharmaceutical industry. These results also validated the previous literature [9,14,15,22,64,125,126].

29. Conclusions

The undertaken research concluded that there are several factors which are significant for the symmetry of physicians' prescription behavior in the context of Pakistani pharmaceutical industry. The results of the study concluded that predictors such as medical representatives' effectiveness, medical literature & journal advertising, scientific activities, personal obligations, promotional material, and direct-to-consumers advertising have positive and significant influence on the physicians' symmetrical prescription behavior. Since the promotional material has the highest positive and significant, the local and transnational pharmaceutical companies should give maximum allocation of budget to the promotional material such as literature on drugs, prescription information monographs, free samples, mountings, and reminder gifts. The scientific activities influence is second; therefore, companies should allocate separate budgets to scientific activities, such as round table discussions, medical group meetings, wards presentations, chamber presentations, and local & foreign speaker programs. Human resources are a vital element for every industry, and especially in pharmaceutical industry, where one-to-one selling takes place; the medical representatives' competency plays a significant role. Thus, the companies should initiate periodic technical and selling skills training programs to enhance the efficacy of MRs effectiveness. The concept of DTCA has been very popular and effective with the advancement of technology; hence, the pharmaceutical companies should also hire a separate team to promote their brands through Internet channels. The undertaken study also concluded that the corporate image and customer relationship are significant moderators that enhance the numbers of prescriptions; therefore, companies should also focus the CSR activities that can increase their corporate image. Moreover, there should be a customized approach to promote their products to every individual physician.

30. Managerial Policy Implications

The outcomes of this research study provide the foundations to the brand/marketing/medical managers of the pharmaceutical industry for how they should respond to physicians' prescription practices. Medical managers should cater the needs of medical literature, current studies, products monographs, and medical journal advertising, which are essentially required by the medical practitioners. The results clearly demonstrated that the medical literature and journal advertising play a vital role in reshaping the symmetry of prescription behavior of physicians; thus, special emphasis should be given while making and producing this literature for physicians. The most important managerial implication of this study is to devise rational and effective marketing strategies according to brand and marketing managers. They should be more vigilant and responsive while making strategies in terms of promotional material, scientific activities, training and development of MRs, personal obligations, and the use of online media for DTCA communication. Since the results have shown these channels have a direct and significant influence on the symmetry of prescription business, marketers should utilize their marketing resources in an effective and efficient manner to gain a competitive advantage. Finally, marketers and medical managers should also devise a program for corporate image building and CRM programs to retain and enhance existing and potential customers.

31. Limitations and Delimitations of the Research

The outcomes of this research have demonstrated the effectiveness of medical, marketing, and moderating variables on how to get the maximum prescription business from physicians. However, nothing could be generalized for every culture and society; factors can be varied for developed and developing economies. Thus, there is a need to examine other marketing and medical factors, which are pertinent to diverse cultures and medical specialties on a regular basis to device effective marketing strategies. Moreover, there are certain mediating and moderating variables that can influence the symmetry of physicians' prescription behavior. So, it is recommended for the future research studies that those factors be incorporated for more robust results. The geographic scope is also very limited; thus, it is also recommended that regional countries replicate this research for a better understanding of physicians' attitude towards the prescription business. We have employed an SEM-based multivariate approach that ignores the causality description; therefore, it is also recommended that future researchers employ causal modeling approaches to measure the influential and causal effects to comprehend the behavioral and psychological intentions of physicians' symmetrical prescription trends.

Author Contributions: Conceptualization, D.S. and R.A.; methodology, J.V., Z.C.; software, Z.C. and R.S.; validation, R.A. and V.P.; formal validation, R.S. and J.V.; writing, R.A. and V.P.; visualization, R.S. and Z.C.; supervision, D.S. and R.A.; project administration, D.S. and Z.C.

Funding: This research received no external funding.

Conflicts of Interest: The authors declare no conflicts of interest.

Disclosure Statement: Evidently, researchers have not reported any impending conflict of interest.

References

1. Ahmed, R.R.; Vveinhardt, J.; Štreimikienė, D.; Ashraf, M.; Channar, Z.A. Modified SERVQUAL Model and Effects of Customer Attitude and Technology on Customer Satisfaction in Banking Industry: Mediation, Moderation and Conditional Process Analysis. *J. Bus. Econ. Manag.* **2017**, *18*, 974–1004. [CrossRef]
2. Ahmed, R.R.; Vveinhardt, J.; Štreimikienė, D.; Awais, M. Mediating and Marketing factors influence the prescription behavior of Physicians: An Empirical Investigation. *Amfiteatru Econ.* **2016**, *18*, 153–167.

3. Ahmed, R.R.; Vveinhardt, J.; Štreimikienė, D. The direct and indirect impact of Pharmaceutical industry in Economic expansion and Job creation: Evidence from Bootstrapping and Normal theory methods. *Amfiteatru* **2018**, *20*, 454–469. [CrossRef]

4. Aikin, K.J.; Swasy, J.L.; Braman, A.C. *Patient and Physician Attitudes and Behaviors Associated with DTC Promotion of Prescription Drugs*; U.S. Department of Health and Human Services Food and Drug Administration: Silver Spring, MD, USA, 2004.

5. Alosaimi, F.D.; Al Kaabba, A.; Qadi, M.; Albahlal, A.; Alabdulkarim, Y.; Alabduljabbar, M.; Alqahtani, F. Physicians' attitudes towards interaction with the pharmaceutical industry/Attitudes des médecins par rapport á l'interaction avec l'industrie pharmaceutique. *East. Med. Health J.* **2014**, *20*, 812–819. [CrossRef]

6. Anderson, E.W.; Sullivan, M.W. The Antecedents and Consequences of Customer Satisfaction for Firms. *Mark. Sci.* **2003**, *12*, 125–143. [CrossRef]

7. Anderson, J.C.; Gerbing, D.W. Structural equation modeling in practice: A review and recommended two-step approach. *Psychol. Bull.* **1988**, *103*, 411–423. [CrossRef]

8. Anon. Study finds problems in M.D.-sales rep relationship. *Pharm Representative*, 6 August 2003.

9. Arora, U.; Taneja, G. An Analytical Study of Physicians Behavior towards Marketing of Pharmaceutical Products. *Indian J. Mark.* **2006**, *36*, 10–13.

10. Avlonitis, G.J.; Panagopoulos, N.G. Antecedents and Consequences of CRM Technology Acceptance in the Sales Force. *Ind. Mark. Manag.* **2004**, *33*, 475–489. [CrossRef]

11. Bala, R.; Bhardwaj, P. Detailing vs. Direct-to-Consumer Advertising in the Prescription Pharmaceutical Industry. *Manag. Sci.* **2010**, *56*, 148–160. [CrossRef]

12. Balmer, J.M.T.; Greyser, S.A. Managing the multiple identities of the corporation. *Calif. Mana. Rev.* **2002**, *44*, 72–86. [CrossRef]

13. Birkhahn, R.H.; Jauch, E.; Kramer, D.A.; Nowak, R.M.; Raja, A.S.; Summers, R.L.; Weber, J.E.; Diercks, D.B. A review of the federal guidelines that inform and influence relationships between physicians and industry. *Acad. Emerg. Med.* **2009**, *16*, 776–781. [CrossRef] [PubMed]

14. Bolin, J.H. Hayes, Andrew, F. (2013). Introduction to Mediation, Moderation, and Conditional Process Analysis: A Regression-Based Approach. New York, NY: The Guilford Press. *J. Educ. Meas.* **2014**, *51*, 335–337. [CrossRef]

15. Boltri, J.M.; Gordon, E.R.; Vogel, R.L. Effect of antihypertensive samples on physician prescribing patterns. *Fam. Med.* **2002**, *34*, 729–731. [PubMed]

16. Bowman, M.A.; Pearle, D.L. Changes in drug prescribing patterns related to commercial company funding of continuing medical education. *J. Contin. Educ. Health Prof.* **1988**, *8*, 13–20. [CrossRef] [PubMed]

17. Brown, S.R.; Evans, D.V.; Fugh-Berman, A. Pharmaceutical industry interactions in family medicine residencies decreased between 2008 and 2013: A CERA study. *Fam. Med.* **2015**, *47*, 279–282.

18. Byrne, B.M. *Structural Equation Modeling with AMOS, Basic Concepts, Application and Programming*, 2nd ed.; La Erlbaum Associates: Hillsdale, NJ, USA, 2009.

19. Campbell, E.G.; Gruen, R.L.; Mountford, J.; Miller, L.G.; Cleary, P.D.; Blumenthal, D. A national survey of Physicians-Industry relationships. *N. Engl. J. Med.* **2007**, *356*, 1742–1750. [CrossRef]

20. Carroll, A.; Barnes, S.J.; Scornavacca, E.; Fletcher, K. Consumer perceptions and attitudes towards SMS advertising: Recent evidence from New Zealand. *Int. J. Advert.* **2007**, *26*, 79–98. [CrossRef]

21. Caudill, T.S.; Johnson, M.S.; Rich, E.C.; McKinney, W.P. Physicians, pharmaceutical representatives, and the cost of prescribing. *Arch. Fam. Med.* **1996**, *5*, 201–206. [CrossRef]

22. Chetley, A. *Problem Drugs*; Health Action International: Amsterdam, the Netherlands, 1995.

23. Chew, L.D.; O'Young, T.S.; Hazlet, T.K.; Bradley, K.A.; Maynard, C.; Lessler, D.S. A physician survey of the effect of drug sample availability on physicians' behavior. *J. Gen. Intern. Med.* **2000**, *15*, 478–483. [CrossRef]

24. Chimonas, S.; Brennan, T.A.; Rothman, D.J. Physicians and Drug Representatives: Exploring the Dynamics of The Relationship. *Soc. Gen. Intern. Med.* **2007**, *22*, 184–190. [CrossRef] [PubMed]

25. Chren, M.M.; Landefeld, C.S. Physicians' behavior and their interactions with drug companies. A controlled study of physicians who requested additions to a hospital drug formulary. *JAMA* **1994**, *271*, 684–689. [CrossRef] [PubMed]

26. Cialdini, R.B. *Influence: Science and Practice*, 3rd ed.; HarperCollins College Publishers: New York, NY, USA, 1993.

27. Clark, M.M.; Gong, M.; Schork, M.; Evans, D.; Roloff, D.; Hurwitz, M.; Maiman, L.; Mellins, R. Impact of Education for Physicians on Patient Outcomes. *Am. J. Pediatr.* **1998**, *101*, 831–836. [CrossRef]

28. Corckburn, J.; Pit, S. Prescribing behaviour in clinical practice: Patients' expectations and doctors' perceptions of patients'expectations—A questionnaire study. *BMJ* **1997**, *315*, 520–523. [CrossRef]

29. Couturier, C.; Zaleski, I.; Jolly, D.; Durieux, P. Effects of financial incentives on medical practice: Results from a systematic review of the literature and methodological issues. *Int. J. Qual. Health Care* **2000**, *12*, 133–142. [CrossRef]

30. Davies, G.; Chun, R.; da Silva, R.V.; Roper, S. *Corporate Reputation and Competitiveness*; Routledge: London, UK, 2003.

31. De Ferrari, A.; Gentille, C.; Davalos, L.; Huayanay, L.; Malaga, G. Attitudes and relationship between physicians and the pharmaceutical industry in a public general hospital in Lima, Peru. *PLoS ONE* **2014**, *9*, e100114. [CrossRef]

32. De Laat, E.; Windmeijer, F.; Douven, R. *How Does Pharmaceutical Marketing Influence Doctors' Prescription Behaviour?* CPB Netherlands' Bureau for Economic Policy Analysis: The Hague, the Netherlands, 2002.

33. DeJong, C.; Aguilar, T.; Tseng, C.W.; Lin, G.A.; Boscardin, W.J.; Dudley, R.A. Pharmaceutical Industry Sponsored Meals and Physician Prescribing Patterns for Medicare Beneficiaries. *JAMA Intern. Med.* **2016**, *176*, 1114–1122. [CrossRef]

34. Dixon, D.; Takhar, J.; Macnab, J.; Eadie, J.; Lockyer, J.; Stenerson, H.; François, J.; Bell, M. Controlling quality in CME/ CPD by measuring and illuminating bias. *J. Contin. Educ. Health Prof.* **2011**, *31*, 109–116. [CrossRef]

35. Dolovich, L.; Levine, M.; Tarajos, R.; Duku, E. Promoting optimal antibiotic therapy for otitis media using commercially sponsored evidence-based detailing: A prospective controlled trial. *Drug Inf. J.* **1999**, *33*, 1067–1077. [CrossRef]

36. Dowling, G.R. *Corporate Reputations. Strategies for Developing the Corporate Brand*; Kogan Page: London, UK, 1994.

37. Dowling, G.R. *Creating Corporate Reputations. Identity, Image and Performance*; Oxford University Press: Oxford, UK, 2001.

38. Dowling, G.R. Corporate reputations: Should you compete on yours? *Calif. Manag. Rev.* **2004**, *46*, 19–36. [CrossRef]

39. Elling, M.E.; Fogle, H.J.; McKhann, C.S.; Simon, C. (2002). Making more of pharma's sales force. *McKinsey Q.* **2002**, *3*, 86–95.

40. Emory, C.W.; Cooper, D.R. *Business Research Methods*, 4th ed.; Irvin: Boston, MA, USA, 1991.

41. Ettenson, R.; Knowles, J. Merging the brands and branding the merger. *Sloan Manag. Rev.* **2006**, *47*, 39–49.

42. Findlay, S. *Prescription Drugs and Mass Media Advertising*; National Institute for Health Care Management: Washington, DC, USA, 2000.

43. Flavian, C.; Torres, E.; Guinaliu, M. Corporate image measurement. A further problem for the tangibilization of Internet banking services. *Int. J. Bank Mark.* **2004**, *22*, 366–384. [CrossRef]

44. Fornell, C.; Larcker, D.F. Evaluating structural equation models with unobservable variables and measurement error. *J. Mark. Res.* **1981**, *18*, 39–50. [CrossRef]

45. Francer, J.; Izquierdo, J.Z.; Music, T.; Narsai, K.; Nikidis, C.; Simmonds, H.; Woods, P. Ethical pharmaceutical promotion and communications worldwide: Codes and regulations. *Philos. Ethics Humanit. Med.* **2014**, *9*, 7. [CrossRef]

46. Gagnon, M.A.; Lexchin, J. The cost of pushing pills: A new estimate of pharmaceutical promotion expenditures in the United States. *PLoS Med.* **2008**, *5*, e1. [CrossRef]

47. Gonül, F.F.; Carter, F.; Petrova, E.; Srinivasan, K. Promotion of Prescription Drugs and its Impact on Physicians' Choice Behavior. *J. Mark.* **2001**, *65*, 79–90. [CrossRef]

48. Gotsi, M.; Wilson, A.M. Corporate reputation: Seeking a definition. *Corp. Commun. Int. J.* **2001**, *6*, 24–30. [CrossRef]

49. Greene, J. Drug Reps. In *Targeting Non-Physicians*; American Medical Association: Chicago, IL, USA, 2000.

50. Greyser, S.A. Advancing and enhancing corporate reputation. *Corp. Commun. J.* **1999**, *4*, 177–181. [CrossRef]

51. Groves, K.E.M.; Sketris, I.; Tett, S.E. Prescription drug samples–Does this marketing strategy counteract policies for quality use of medicines? *J. Clin. Pharm. Ther.* **2003**, *28*, 259–271. [CrossRef]

52. Grundy, Q.; Bero, L.; Malone, R. Interactions between non-physician clinicians and industry: A systematic review. *PLoS Med.* **2013**, *10*, e1001561. [CrossRef] [PubMed]

53. Gupta, S.K.; Nayak, R.P.; Sivaranjani, R. A study on the interactions of doctors with medical representatives of pharmaceutical companies in a Tertiary Care Teaching Hospital of South India. *J. Pharm. Bioallied Sci.* **2016**, *8*, 47–51. [CrossRef] [PubMed]

54. Haddad, J.K. The Pharmaceutical Industry's Influence on Physician Behavior and Health Care Costs. 2002. Available online: https://www.sfms.org/sfm/sfm602a.htm (accessed on 29 May 2003).
55. Hair, J.F.; Black, W.C.; Babin, B.J.; Anderson, R.E. *Multivariate Data Analysis*, 7th ed.; Prentice Hall: Upper Saddle River, NJ, USA, 2010.
56. Hall, R. The strategic analysis of intangible resorces. *Strateg. Manag. J.* **1992**, *13*, 135–144. [CrossRef]
57. Halperin, E.C.; Hutchison, P.; Barrier, R., Jr. A population-based study of the prevalence and influence of gifts to radiation oncologists from pharmaceutical companies and medical equipment manufacturers. *Int. J. Radiat. Oncol. Biol. Phys.* **2004**, *59*, 1477–1483. [CrossRef] [PubMed]
58. Hayes, A.F. *Introduction to Mediation, Moderation, and Conditional Process Analysis: A Regression-Based Approach*; The Guilford Press: New York, NY, USA, 2013.
59. Hersh, W.R.; Crabtree, K.M.; Hickam, D.H.; Sacherek, Y.; Friedman, C.P.; Tidmarch, P.; Mosbaek, C.; Kramier, D. Factors associated with success in searching MEDLINE and applying evidence to answer clinical questions. *J. Am. Med. Informat. Assoc.* **2002**, *9*, 283–293. [CrossRef]
60. Herzenstein, M.; Misra, S.; Posavac, S.S. How consumers' attitudes toward direct-to-consumer advertising of prescription drugs influence ad effectiveness, and consumer and physician behavior. *Mark. Lett.* **2004**, *15*, 201–212. [CrossRef]
61. Holmer, A.F. Industry strongly supports continuing medical education. *JAMA* **2001**, *285*, 2012–2014. [CrossRef] [PubMed]
62. Huang, J.H.; Lee, B.C.Y.; Ho, S.H. Consumer attitude toward grey market goods. *Int. Mark. Rev.* **2004**, *21*, 598–614. [CrossRef]
63. Hurley, M.P.; Stafford, R.S.; Lane, A.T. Characterizing the relationship between free drug samples and prescription patterns for acne vulgaris and rosacea. *JAMA Dermatol.* **2014**, *150*, 487–493. [CrossRef]
64. Jaffe, E.D.; Nebenzahl, I.D. *National Image & Competitive Advantage, the Theory and Practice of Place Branding*, 2nd ed.; Copenhagen Business School Press: Frederiksberg, Denmark, 2006.
65. Kaiser, H.F. An index of factorial simplicity. *Psychometrika* **1974**, *39*, 31–36. [CrossRef]
66. Kamal, S.; Holmberg, C.; Russell, J.; Bochenek, T.; Tobiasz-Adamczyk, B.; Fischer, C.; Tinnemann, P. Perceptions and Attitudes of Egyptian Health Professionals and Policy-Makers towards Pharmaceutical Sales Representatives and Other Promotional Activities. *PLoS ONE* **2015**, *10*, e0140457. [CrossRef] [PubMed]
67. Katz, D.; Caplan, A.L.; Merz, J.F. All gifts large and small. *Am. J. Bioethics* **2003**, *2*, 39–46. [CrossRef] [PubMed]
68. Kesselheim, A.S.; Mello, M.M.; Studdert, D.M. Strategies and practices in off label marketing of pharmaceuticals: A retrospective analysis of whistleblower complaints. *PLoS Med.* **2011**, *8*, e1000431. [CrossRef] [PubMed]
69. Khan, N.; Naqvi, A.; Ahmad, R.; Ahmed, F.; McGarry, K.; Fazlani, R.; Ahsan, M. Perceptions and Attitudes of Medical Sales Representatives (MSRs) and Prescribers Regarding Pharmaceutical Sales Promotion and Prescribing Practices in Pakistan. *J. Young Pharm.* **2016**, *8*, 244–250. [CrossRef]
70. Leeflang, P.S.H.; Wieringa, J.E. Modeling the effects of pharmaceutical marketing. *Mark. Lett.* **2010**, *21*, 121–133. [CrossRef]
71. Leffler, K.B. Persuasion or Information? The Economics of Prescription Drug Advertising. *J. Law Econ.* **1981**, *24*, 45–74. [CrossRef]
72. Lexchin, J. Interactions between physicians and the pharmaceutical industry: What does the literature say? *Can. Med. Assoc. J.* **1993**, *14*, 1401–1407.
73. Lieb, K.; Koch, C. Medical students' attitudes to and contact with the pharmaceutical industry: A survey at eight German university hospitals. *Dtsch. Arztebl. Int.* **2013**, *110*, 584–590.
74. Lieb, K.; Scheurich, A. Contact between Doctors and the Pharmaceutical Industry, Their Perceptions, and the Effects on Prescribing Habits. *PLoS ONE* **2014**, *9*, e110130. [CrossRef] [PubMed]
75. Liebman, M. Rethinking sampling strategies. *Med. Mark. Media* **1997**, *32*, 36–43.
76. Liu, Q.; Gupta, S. The impact of direct-to-consumer advertising of prescription drugs on physician visits and drug request: Empirical findings and public policy implications. *Int. J. Res. Mark.* **2011**, *28*, 205–217. [CrossRef]
77. Loden, D.J.; Liebman, M. High Compression Marketing Supercharges Drug Revenues. 2000. Available online: https://www.mmm-online.com/mmm-magazine/ (accessed on 21 March 2001).
78. Lotfi, T.; Morsi, R.Z.; Rajabbik, M.H.; Alkhaled, L.; Kahale, L.; Hala, N.; Brax, H.; Fadiallah, R.; Aki, E.A. Knowledge, beliefs and attitudes of physicians in low and middle-income countries regarding interacting with pharmaceutical companies: A systematic review. *BMC Health Serv. Res.* **2016**, *16*, 57. [CrossRef] [PubMed]

79. Lurie, N.; Rich, E.C.; Simpson, D.E.; Meyer, J.; Schiedermayer, D.L.; Goodman, J.L.; McKinney, W.P. Pharmaceutical representatives in academic medical centers. *J. Gen. Intern. Med.* **1990**, *5*, 240–243. [CrossRef] [PubMed]

80. MacCallum, R.C.; Widaman, K.; Zhang, S.; Hong, S. Sample size in Factor Analysis. *Psychol. Methods* **1999**, *4*, 84–99. [CrossRef]

81. Makowska, M. Interactions between doctors and pharmaceutical sales representatives in a former communist country, the ethical issues. *Camb. Q. Healthc. Ethics* **2014**, *23*, 349–355. [CrossRef]

82. Manchanda, P.; Phil, M.; Honka, E. The Effects and Role of Direct-to-Physician Marketing in the Pharmaceutical Industry: An Integrative Review. *Yale J. Health Policy Law Ethics* **2005**, *5*, 785–822.

83. Marco, C.A.; Moskop, J.C.; Solomon, R.C.; Geiderman, J.M.; Larkin, G.L. Gifts to Physicians from the Pharmaceutical Industry: An Ethical Analysis. *Ann. Emerg. Med.* **2006**, *48*, 513–521. [CrossRef] [PubMed]

84. Markwick, N.; Fill, C. Towards a framework for managing corporate identity. *Eur. J. Mark.* **1997**, *31*, 396–410. [CrossRef]

85. Masood, I.; Ibrahim, M.; Hassali, M.; Ahmad, M. Evolution of marketing techniques, Adoption in Pharmaceutical Industry and related issues: A review. *J. Clin. Diagn. Res.* **2009**, *7*, 1942–1952.

86. Mehta, P.N. Drug makers and continuing medical education. *Indian Pediatr.* **2000**, *37*, 626–630.

87. Mizik, N.; Jacobson, R. Are physicians "easy marks"? Quantifying the effects of detailing and sampling on new prescriptions. *Manag. Sci.* **2004**, *50*, 1704–1715. [CrossRef]

88. Montastruc, F.; Moulis, G.; Palmaro, A.; Gardette, V.; Durrieu, G.; Montastruc, J.-L. Interactions between medical residents and drug companies: A national survey after the Mediator® affair. *PLoS ONE* **2014**, *9*, e104828. [CrossRef] [PubMed]

89. Mukherji, P. Estimating the effects of direct-to-consumer advertising for prescription medicines: A natural experiment. In Proceedings of the Marketing Science Conference, Rotterdam, NL, USA, 24–26 June 2004.

90. Mulinari, S. Unhealthy marketing of pharmaceutical products: An international public health concern. *J. Public Health Policy* **2016**, *37*, 149–159. [CrossRef] [PubMed]

91. Nair, H.S.; Manchanda, P.; Bhatia, T. Asymmetric social interactions in physician prescription behavior: The role of opinion leaders. *J. Mark. Res.* **2010**, *47*, 883–895. [CrossRef]

92. Narayanan, S.; Desiraju, R.; Chintagunta, P.K. Return on Investment Implications for Pharmaceutical Promotional Expenditures: The Role of Marketing-Mix Interactions. *J. Mark.* **2004**, *68*, 90–105. [CrossRef]

93. Newell, S.J.; Goldsmith, R.E. The development of a scale to measure perceived corporate credibility. *J. Bus. Res.* **2001**, *52*, 235–247. [CrossRef]

94. Orlowski, J.P.; Wateska, L. The effects of pharmaceutical firm enticements on physician prescribing patterns. There's no such thing as a free lunch. *Chest* **1992**, *102*, 270–273. [CrossRef]

95. Osinga, E.C.; Leeflang, P.S.; Wieringa, J.E. Early Marketing Matters: A Time-Varying Parameter Approach to Persistence. *J. Mark. Res.* **2010**, *47*, 173–185. [CrossRef]

96. Othman, N.; Vitry, A.I.; Roughead, E.E.; Ismail, S.B.; Omar, K. Medicines information provided by pharmaceutical representatives: A comparative study in Australia and Malaysia. *BMC Public Health* **2010**, *10*, 743. [CrossRef]

97. Parikh, K.; Fleischman, W.; Agrawal, S. Industry relationships with pediatricians: Findings from the open payments sunshine act. *Pediatrics* **2016**, *137*, e20154440. [CrossRef]

98. Peppin, J.F. Pharmaceutical sales representatives and physicians: Ethical considerations of a relationship. *J. Med. Philos.* **1996**, *21*, 83–99. [CrossRef] [PubMed]

99. Pinto, C.A.S.O.; Green, D.; Baby, A.R.; Ruas, G.W.; Kaneko, T.M.; Marana, S.R.; Velasco, M.V.R. Determination of papain activity in topical dosage forms: Single laboratory validation assay. *Lat. Am. J. Pharm.* **2007**, *26*, 771–775.

100. Pirisi, A. Patient-direct drug advertising puts pressure on US doctors. *Lancet* **1999**, *354*, 1887. [CrossRef]

101. Rahman, M.H.; Agarwal, S.; Tuddenham, S.; Peto, H.; Iqbal, M.; Bhuiya, A.; Peters, D.H. What do they do? Interactions between village doctors and medical representatives in Chakaria, Bangladesh. *Int. Health* **2015**, *7*, 266–271. [CrossRef] [PubMed]

102. Relman, A.S. Separating continuing medical education from pharmaceutical marketing. *JAMA* **2001**, *285*, 2009–2012. [CrossRef]

103. Rizzo, J. Advertising and competition in the ethical pharmaceutical industry: The case of antihypertensive drugs. *J. Law Econ.* **1999**, *42*, 89–116. [CrossRef]

104. Rogers, E. *Diffusion of Innovations*; Free Press: New York, NY, USA, 1995.

105. Rutherford, G.S.W.; Hair, J.F.; Anderson, R.E.; Tatham, R.L. Multivariate data analysis. *Statistician* **1988**, *37*, 484. [CrossRef]

106. Saito, S.; Mukohara, K.; Bito, S. Japanese practicing physicians' relationships with pharmaceutical representatives: A national survey. *PLoS ONE* **2010**, *5*, e12193. [CrossRef]

107. Scheffer, M.C. Interaction between pharmaceutical companies and physicians who prescribe antiretroviral drugs for treating AIDS. *Sao Paulo Med. J.* **2014**, *132*, 55–60. [CrossRef]

108. Schwartzman, D. *Innovation in the Pharmaceutical Industry*; Johns Hopkins University Press: Baltimore, MD, USA, 1976.

109. Shalowitz, D.I.; Spillman, M.A.; Morgan, M.A. Interactions with industry under the Sunshine Act: An example from gynecologic oncology. *Am. J. Obstet. Gynecol.* **2016**, *214*, 703–707. [CrossRef]

110. Siddiqui, U.T.; Shakoor, A.; Kiani, S.; Ali, F.; Sharif, M.; Kumar, A.; Raza, Q.; Khan, N.; Alamzaib, S.M.; Farid-ul-Husnain, S. Attitudes of medical students towards incentives offered by pharmaceutical companies—Perspective from a developing nation—A cross-sectional study. *BMC Med. Ethics* **2014**, *15*, 36. [CrossRef] [PubMed]

111. Smith, A.D. Corporate Social Responsibility Practices in the Pharmaceutical Industry. *Bus. Strategy Ser.* **2008**, *9*, 306–315. [CrossRef]

112. Spingarn, R.W.; Berlin, J.A.; Strom, B.L. When pharmaceutical manufacturers' employees present grand rounds, what do residents remember? *Acad. Med.* **1996**, *71*, 86–88. [CrossRef] [PubMed]

113. Stafford, R.S.; Furberg, C.D.; Finkelstein, S.N.; Cockburn, I.M.; Alehegn, T. Impact of clinical trial results on national trends in alpha-blocker prescribing, 1996–2002. *JAMA* **2004**, *291*, 54–62. [CrossRef] [PubMed]

114. Stros, M.; Hari, J.; Marriott, J. The Relevance of Marketing Activities in The Swiss Prescribing Drugs Market: Two Empirical Studies. *Int. J. Pharm. Healthc. Mark.* **2009**, *3*, 323–346. [CrossRef]

115. Taneja, G. Influence of Promotional Tools Offered by Pharmaceutical Industry on Physicians Prescribing Behavior. 2007. Available online: https://ssrn.com/abstract=2049747 (accessed on 22 September 2010).

116. Van den Bosch, C.; de Jong, M. How corporate visual identity supports reputation. *Corp. Commun. Int. J.* **2005**, *10*, 108–116. [CrossRef]

117. Van den Bulte, C.; Lilien, G.L. Medical Innovation Revisited: Social Contagion versus Marketing Effort. *Am. J. Sociol.* **2001**, *106*, 1409–1435. [CrossRef]

118. Vicciardo, L.J. The secret weapon in marketing intelligence: Meetings and events analysis is a powerful new tool in researching marketing strategy and success. *Med. Mark. Media* **1995**, *30*, 32–36.

119. Watkins, C.; Moore, L.; Harvey, I.; Carthy, P.; Robinson, E.; Brawn, R. Attitudes and Behavior of General Practitioners and Their Prescribing Costs: A National Cross Sectional Survey. *Qual. Saf. Health Care* **2003**, *12*, 29–34. [CrossRef]

120. Wazana, A. Doctors and the Pharmaceutical Industry Is A Gift Ever Just A Gift? *J. Am. Med. Assoc.* **2000**, *283*, 373–380. [CrossRef]

121. Wilkes, M.S.; Doblin, B.H.; Shapiro, M.F. Pharmaceutical advertisements in leading medical journals: Experts' assessments. *Ann. Intern. Med.* **1992**, *116*, 912–919. [CrossRef] [PubMed]

122. Windmeijer, F.; De Laat, E.; Douven, R.; Mot, E. Pharmaceutical promotion and GP prescription behaviour. *Health Econ.* **2005**, *15*, 5–18. [CrossRef] [PubMed]

123. Worcester, R. Corporate Image. In *Consumer Market Research Handbook*; Worcester, R., Downham, J., Eds.; McGraw Hill: London, UK, 1986; pp. 601–606.

124. Yeh, J.S.; Austad, K.E.; Franklin, J.M.; Chimonas, S.; Campbell, E.G.; Avorn, J.; Kesselheim, A.S. Association of medical students' reports of interactions with the pharmaceutical and medical device industries and medical school policies and characteristics: A cross-sectional study. *PLoS Med.* **2014**, *11*, e1001743. [CrossRef]

125. Yeh, J.S.; Franklin, J.M.; Avorn, J.; Landon, J.; Kesselheim, A.S. Association of Industry Payments to Physicians With the Prescribing of Brand name Statins in Massachusetts. *JAMA Intern. Med.* **2016**, *176*, 763–768. [CrossRef] [PubMed]

126. Ziegler, M.G. The Accuracy of Drug Information from Pharmaceutical Sales Representatives. *J. Am. Med. Assoc.* **1995**, *273*, 1296–1298. [CrossRef]

MDPI
St. Alban-Anlage 66
4052 Basel
Switzerland
Tel. +41 61 683 77 34
Fax +41 61 302 89 18
www.mdpi.com

Symmetry Editorial Office
E-mail: symmetry@mdpi.com
www.mdpi.com/journal/symmetry